BRICKWORK

BOOKS IN THE REVISION AND SELF-ASSESSMENT SERIES FROM LEEDS COLLEGE OF BUILDING

Ideal for the student working alone, these new books allow readers to test their understanding of key subjects. Each topic begins with a summary of key facts and figures and follows with multiple-choice assessments. Students find these books stimulating and useful revision aids.

- Contains reference notes and definitions
- Includes multiple-choice questions
- Allows students to assess their progress
- Contains notes for revision

PUBLISHED
BRICKWORK
ISBN 0 7506 5072 9

PLUMBING
ISBN 0 340 71911 7

FORTHCOMING
CARPENTRY AND JOINERY
ISBN 0 7506 5116 4

PAINTING AND DECORATING
ISBN 0 7506 5117 2

Also in the revision and self-assessment series

ELECTRICAL INSTALLATION
THEORY AND PRACTICE
Maurice Lewis
ISBN 0 340 67665 5

ELECTRONIC PRINCIPLES AND APPLICATIONS
John B. Pratley
ISBN 0 340 69275 8

ENGINEERING DRAWING FROM FIRST PRINCIPLES USING AUTOCAD
Dennis Maguire
ISBN 0 340 69198 0

Revision and Self-Assessment Series

BRICKWORK

JOHN CARRUTHERS
on behalf of the Leeds College of Building

OXFORD AUCKLAND BOSTON JOHANNESBURG MELBORNE NEW DELHI

Butterworth-Heinemann
Linacre House, Jordan Hill, Oxford OX2 8DP
225 Wildwood Avenue, Woburn, MA 01801-2041
A division of Reed Educational and Professional Publishing Ltd

℞ A member of the Reed Elsevier plc group

First published 2000

British Library Cataloguing in Publication Data
Carruthers, John
Brickwork – Revision and self-assessment series
1. Bricklaying   2. Bricklaying, Problems, exercises, etc.
I. Title   II. Leeds College of Building
693.2'1

ISBN 0 7506 5072 9

Typeset by Phoenix Photosetting, Chatham, Kent
Printed and bound in Great Britain by MPG Books Ltd, Bodmin, Cornwall

Cover photograph: British Telecom, Brindley Place, Birmingham. Reproduced by kind
permission of the Brick Development Association. Copyright: Martine Hamilton Knight

CONTENTS

ACKNOWLEDGEMENTS

Many thanks to the authors of the following publications for permission to use their artwork:

The BDA Guide to Successful Brickwork, 2nd edition, Arnold 2000

Kevin Stead on behalf of the Leeds College of Building, *Plumbing*, Arnold 1999

John Hodge and Bob Baldwin, *Brickwork for Apprentices*, 4th edition, Arnold, 1993

Cover Photograph: British Telecom, Brindley Place, Birmingham. Reproduced by kind permission of the Brick Development Association. Copyright: Martine Hamilton Knight.

1

CONTRIBUTING TO THE PROVISION OF A HEALTHY AND SAFE WORKING ENVIRONMENT

To tackle the assessments in this chapter you will need to know:

- the purpose and scope of the statutory and non-statutory regulations governing the control of health and safety in the construction industry;
- the employers' and employees' duties and responsibilities regarding health and safety in the construction industry;
- the correct procedures in reporting an accident;
- the correct procedures for treatment of accidents;
- the correct procedures in dealing with fires and emergency evacuation procedures;
- the employees' responsibilities towards the customer and general behaviour expected on site.

GLOSSARY OF TERMS

Accident – an event causing injury or damage that could have been avoided by following the correct methods and procedures.

Appointed person – a person authorized by an employer to take care of first aid arrangements.

Competent person – a person who has the experience and technical knowledge to carry out specific tasks.

Control of Substances Hazardous to Health Regulations 1994 (COSHH) – regulations aimed at protecting individuals against risk to their health arising from exposure to hazardous substances.

First aider – a person who is trained in first aid.

Health and Safety at Work Act 1974 (HASAWA) – the main statutory legislation covering health and safety of all persons at their workplace.

A selection of relevant health and safety legislation used in the construction industry

Legislation	Regulations
Control of Pollution Act 1974	Construction (Health, Safety and Welfare) Regulations 1996
Explosives Act 1923	
Factories Act 1961	Provision of Use of Work Equipment Regulations 1998
	Control of Asbestos at Work (amendment) Regulations 1998
	Construction (Health, Safety and Welfare) Regulations 1996
	Lifting Equipment and Lifting Operations Regulations 1998
	Construction (Head Protection) Regulations 1989
	Highly Flammable Liquids and Liquefied Petroleum Gases Regulations 1972
	Control of Lead at Work Regulations 1998
	Protection of Eyes Regulations 1974
	Work in Compressed Air Special Regulations 1996
Fire Precautions Act 1971	Fire Certificates (Special Premises) Regulations 1976

Health and Safety at Work Act 1974	Hazardous Substances (Labelling of Road Tankers) Regulations 1978
	Control of Lead at Work Regulations 1998
	Health and Safety (Safety signs and Signals) Regulations 1995
	Heath and Safety (First Aid) Regulations 1981
	Control of Asbestos at Work Regulations 1998
	Control of Substances Hazardous to Health Regulations 1999
	Reporting of Injuries, Diseases and Dangerous Occurrence's Regulations 1995 (RIDDOR)
Mines and Quarries Act 1954	Health and Safety (Young Persons) Regulations 1997
Offices, Shops and Railway Premises Act 1963	Health and Safety (First Aid) Regulations 1981

Prohibition notice – a document served on an employer who has contravened a legal requirement under the Health and Safety at Work Act 1974.

Safety policy – a written legal statement by an employer who employs five or more people which sets out the general policy of health and safety.

SAFETY SIGNS

Prohibition – shows what **must not be done**.

Mandatory – shows what must be done.

Warning – shows warning of a hazard or danger.

Information – shows information regarding the provision for safety.

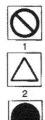

Category
1. Prohibition
2. Warning
3. Mandatory
   first aid/
   safe condition
4. Emergency/escape/
   first aid/
   safe condition
5. Fire equipment

Meaning
1. Do not do
2. Danger
3. Must do
4. The safe way
5. Location or use of fire equipment

Note
Prohibition sign – is a red circle on a white background
Warning sign – is a yellow triangle with a black border on a white background
Mandatory sign – is a blue square/rectangle on a white background
Emergency sign – is a green square/rectangle on a white background
Fire equipment – is a red square/rectangle on a white background

Saftey signs

Assessments 1.1, 1.2 and 1.3

CONTRIBUTING TO THE PROVISION OF A HEALTHY AND SAFE WORKING ENVIRONMENT

Time allowed

Section 1.1: 1 hour
Section 1.2: 1 hour
Section 1.3: 1½ hours

Instructions

- You will need to have the following:
  Question paper
  Answer sheet
  Pencil.
- Ensure your name and date is at the top of the answer sheet.
- When you have decided a correct answer, draw a straight line through the appropriate letter on the answer sheet.
- If you make a mistake with your answer, change the original line by making it into a cross and then put a line through the amended answer. There is only one answer to each question.
- Do not write on the question sheet.
- Make sure you read each question carefully and try to answer all the questions in the time allowed.

Example

a	150 mm	~~**a**~~	150 mm
b	75 mm	**b**	75 mm
~~**c**~~	225 mm	✗**c**	225 mm
d	300 mm	**d**	300 mm

Section 1.1

1. Which one of the following Acts of Parliament addresses all people at work and imposes duties on everyone?
   a Construction (Working) Places Regulations
   b Health and Safety and Work Act 1974
   c Control of Substances Hazardous to Health Regulations 1989
   d Mines and Quarries Act 1954.

2. Which one of the following is not a function of the safety officer?
   a investigating dangerous occurrences
   b investigating complaints by employees
   c carrying out work inspections
   d writing reports for HSE.

3. What is the meaning of the mandatory safety sign shown below?

   a wear head protection
   b wear hearing protection
   c wear hand protection
   d wear foot protection.

4. Which of the following tools is likely to develop a mushroom head?
   a walling trowel
   b bolster chisel
   c spirit level
   d brick hammer.

5. Which one of the following safety signs below could be applicable in a work situation involving cutting bricks?
   a
   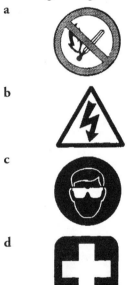
   b

   c

   d

6. The symbol shown below is used for labelling substances that are:

   a irritant
   b solvents
   c harmful
   d corrosive.

7. All accidents must be reported to the HSE if the person has been off work for more than:
   a 2 days
   b 3 days
   c 4 days
   d 5 days.

8. A prohibition sign is:
   a   circular with a red border and cross bar
   b   circular with a white symbol on a blue background
   c   triangular with a yellow background with a black border
   d   square or rectangular with a white symbol on a green background.

9. A mandatory sign is one that means:
   a   must not be done
   b   gives information
   c   warning sign
   d   must be done.

10. If you accidentally cause damage to an electric drill, what should you do?
    a   try and repair it yourself
    b   report it to your supervisor
    c   ignore it
    d   leave it for someone else to discover.

11. Which one of the following signs should be displayed when working on a fragile roof?

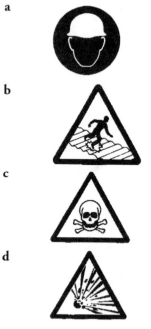

a

b

c

d

12. Which one of the following safety protection items is required when using a comb hammer?
    a   goggles
    b   helmet
    c   gloves
    d   boots.

13. The symbol shown below is used for labelling substances that are:

    a toxic
    b corrosive
    c irritant
    d non-poisonous.

14. What is the type of safety sign shown below?

    a warning
    b mandatory
    c prohibition
    d information.

15. If an employer employs five or more staff, what should he provide?
    a a safety statement
    b a register of accidents
    c a permanent safety officer
    d provide first aid personnel.

16. Which safety sign below is used for giving a general warning against risk of explosive materials?

    a

    b

    c

    d

17. The meaning of the safety sign below is:

    a wear a hard hat
    b wear safety goggles
    c no smoking or naked flame
    d scaffolding incomplete.

18. The safety sign comprising a white cross on a green background is used for giving information on:
    a first aid
    b dangerous crossing
    c emergency fire escape
    d safe working conditions.

19. Which one of the following is least likely to cause dermatitis?
    a cement
    b tar
    c plaster
    d barrier cream

20. Which one of the following hazard symbols shown below is used for labelling substances that are toxic?

    a

    b

    c

    d

21. You can reduce the risk of electric shock by reducing the voltage on building sites to:
    a 240 V
    b 110 V
    c 230 V
    d 50 V.

22. Which person can issue a prohibition notice on an employer?
    a building inspector
    b clerk of works
    c health and safety inspector
    d architect.

23. When a person starts work for the first time they should be made aware of the:
   a   safety policy
   b   working hours
   c   home time
   d   overtime rates.

24. When lifting something, you should avoid injury to yourself by:
   a   using your arms
   b   keeping your feet on the ground
   c   using your leg muscles
   d   using your back muscles.

25. What does PPE stand for?
   a   personnel protective equipment
   b   private practicing engineer
   c   personnel protective eye glass
   d   private practicing electrician.

Now check your answers from the grid

Q 1; b	Q 6; d	Q 11; b	Q 16; d	Q 21; b
Q 2; d	Q 7; b	Q 12; a	Q 17; c	Q 22; c
Q 3; c	Q 8; a	Q 13; c	Q 18; a	Q 23; a
Q 4; b	Q 9; d	Q 14; d	Q 19; d	Q 24; b
Q 5; c	Q 10; b	Q 15; a	Q 20; c	Q 25; a

Section 1.2

1. Emergency first aid treatment is a procedure which does not include:
   a   preservation of life
   b   preventing the condition getting worse
   c   promoting recovery
   d   sending the casualty home.

2. The object of placing an unconscious person into the recovery position is to:
   a   reduce the possibility of shock
   b   encourage the heart to beat
   c   help maintain an open airway
   d   make the victim sick.

3. Which item below is not essential in a first aid box?
   a   hand wash
   b   triangular bandage
   c   eye pad
   d   sterile dressing.

4. Which is not part of the triangle of fire?
   a heat
   b light
   c oxygen
   d fuel.

5. Which will not spread fire?
   a convection
   b conduction
   c radiation
   d compartmentalization.

6. What extinguisher would you not use for a burning liquid fire?
   a water
   b foam
   c carbon dioxide
   d dry powder.

7. Which extinguisher would you use for an electric fire?
   a water
   b carbon dioxide
   c foam
   d dry power.

8. What extinguisher would you not use for an oil-waste fire?
   a foam
   b dry powder
   c water
   d carbon dioxide.

9. You can test that a room is involved in a fire by:
   a feeling the door with the palm of your hand
   b feeling the door with the back of your hand
   c feeling the door with the tips of your fingers
   d opening the door.

10. What is one thing that you can do to stop a fire spreading?
    a close the curtains
    b close the works
    c close the door
    d close for lunch.

11. When dealing with a fire, what is the first thing you should do?
    a run away
    b sound the alarm
    c pretend you have not seen it
    d panic.

12. When exiting from the building where should you go?
   a   the pub
   b   back inside to get something
   c   home
   d   a nominated assembly point.

13. Which one of the following should not be considered if someone is suffering from shock?
   a   give something to drink
   b   keep warm
   c   loosen clothing around the neck
   d   treat any injuries.

14. If someone spills some chemical cleaning acid on their hand, what should you do?
   a   burst the blisters
   b   apply ointments
   c   immerse in running water
   d   remove clothing stuck to the wound.

15. What should you do if someone has a badly cut finger?
   a   seek assistance from a first aider
   b   place the wound under running water
   c   send for an ambulance
   d   apply direct pressure to the wound.

16. In the figure below, tilting the head back to begin artificial respiration:
   a   allows more blood to circulate to the brain
   b   allows the casualty to see the rescuer
   c   opens the victim's airway
   d   lifts the victims tongue off the back of his throat.

17. Which one of the following types of fire should a Class D fire extinguisher be must suitable
   for?
   a   metal
   b   plastic
   c   electric
   d   oil.

18. Which of the following extinguishers cannot be used on electric fires?

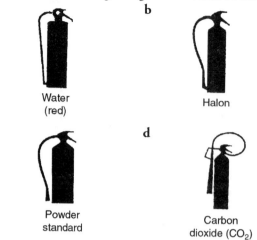

a

Water
(red)

b

Halon

c

Powder
standard

d

Carbon
dioxide (CO_2)

19. In the absence of a first aider, which of the following persons is likely to be asked to treat a casualty?
   a supervisor
   b visiting nurse
   c appointed person
   d passer by.

20. If a colleague at work collapses in a room full of toxic fumes, what should you do?
   a seek medical help
   b open all window and doors
   c apply artificial respiration
   d leave the danger area.

21. The type of extinguisher shown below is used on electric fires because it contains:
   a water
   b foam
   c dry powder
   d CO_2.

22. A carbon dioxide fire extinguisher works by:
   a knocking the flames down with pressure
   b smothering the flames by starving them of air
   c cooling the flames with its water
   d starving the flames of fuel.

23. Which one of the following is not an example of good posture when lifting a load?
   a keeping the spine straight
   b keeping the shoulders level
   c keeping the feet slightly apart
   d keeping the heels off the ground.

24. When team lifting what considerations should you take?
    a   one person should take charge and give all instructions
    b   everyone lifts separately
    c   let someone else take more of the load
    d   three people stand at one side with one at the other.

25. Which one of the following is regarded as a load, as defined by the Manual Handling Regulations 1992?
    a   a saw cutting through metal
    b   a crowbar removing a nail
    c   using a wheelbarrow to move a load
    d   climbing a ladder.

Now check your answers from the grid

Q 1; d	Q 6; a	Q 11; b	Q 16; c	Q 21; d
Q 2; c	Q 7; b	Q 12; d	Q 17; a	Q 22; b
Q 3; a	Q 8; c	Q 13; a	Q 18; a	Q 23; d
Q 4; b	Q 9; b	Q 14; c	Q 19; c	Q 24; a
Q 5; d	Q 10; c	Q 15; d	Q 20; b	Q 25; c

Section 1.3

1. Explain why accidents happen.

2. Define the term 'accident'.

3. What is meant by a reported accident?

4. List five reasons why accidents may occur.

5. State three powers of the health and safety inspector.

6. What is the role of the safety officer?

7. What do the initials COSHH stand for?

8. What is the main purpose of COSHH?

9. What are the main implications of COSHH?

10. What do the initials HASAWA mean?

11. List two objectives of HASAWA.

12. Describe three duties of the employer under HASAWA.

13. Describe three duties of the employee under HASAWA.

14. What do the initials PPE stand for?

15. Where would you use the following protective equipment?
    dust mask/respirator
    safety goggles
    safety helmet.

16. State the type of safety equipment that you should use when handling glass.

17. What are your responsibilities regarding safe working on site?

18. List the shape and colour of the following types of safety signs:
    prohibition
    mandatory
    warning
    information.

19. What do the following symbols mean?

20. Why should you not use drugs or drink alcohol when you are at work?

21. What are the priorities of first aid?

22. What procedures should you take if first aid is needed?

23. What type of fire extinguisher would you use to put out the following types of fire?
    electrical
    adhesives
    propane.

24. Describe the procedures that you should follow if you find a fire.

25. What considerations should you take when lifting something heavy?

26. What are the basic rules for team lifting?

27. State the action to be taken to prevent fires.

28. Give three reasons for keeping your work area tidy.

29. Give four building-site operations where you would insist on wearing PPE?

30. List the rules of manual handling.

Model answers for Section 1.3

1. Explain why accidents happen.
    Accidents occur through people failing to behave responsibly on site.

2. Define the term 'accident'.
    A chance event or an unintentional act resulting in injury or damage that could have been avoided if the correct procedures had been in place.

3. What is meant by a reported accident?
    All accidents should be recorded in the accident book, but accidents that result in death, injury or more than three days off work should be reported to the Health and Safety Executive.

4. List five reasons why accidents may occur.
    lack of concentration
    carelessness
    ignorance
    taking shortcuts
    using the wrong tools/equipment.

5. State three powers of the health and safety inspector.

    Any three of the following:
    enter premises in order to carry out investigations
    take statements
    check records
    give information/advice
    demand the seizure, dismantle or destroy any dangerous equipment/machinery
    issue improvement notices
    issue prohibition notices
    prosecute.

6. What is the role of the safety officer?
   To advise the management on the implementation of the company safety policy and help protect employees and the general public from injury on site.

7. What do the initials COSHH stand for?
   Control of Substances Hazardous to Health Regulations 1999.

8. What is the main purpose of COSHH?
   These regulations are aimed at protecting individuals against the risk to their health arising from exposure to substances hazardous to health in the workplace.

9. What are the main implications of COSHH?
   Employers will need to make suitable assessment of any risks.
   Every employer is obliged to ensure that the exposure of the employee is either prevented or controlled.
   The employer must provide the employee with information on the potential hazards to health.

10. What do the initials HASAWA mean?
    Health and Safety at Work Act 1974.

11. List two objectives of HASAWA.

    Any of the following:
    To secure the health, safety and welfare of all persons at work.
    To protect the general public from risk of health and safety arising out of work.
    To control the use, handling, storage and transporting of explosives and highly flammable substances.
    To control the release of noxious or offensive substances into the atmosphere.

12. Describe three duties of the employer under HASAWA.

    Any of the following:
    Provide and maintain a safe working environment.
    Ensure safe access to and from the workplace.
    Provide and maintain safe machinery, equipment and methods of work.
    Ensure safe handling, transporting and storage of machinery, equipment and materials.
    Provide the employees with the necessary information, instructions, training and supervision to ensure safe working.
    Prepare, update as required and issue to employees a written statement of the firm's safety policy.
    Involve trade unions in all safety matters.

13. Describe three duties of the employee under HASAWA.

    Any of the following:
    Take care at all times and ensure that they do not put themselves, their colleagues or any other persons at risk by their actions.
    Co-operate with their employer to enable them to fulfill the employer's health and safety duties.

Use the equipment and safeguards provided by the employer.
Never misuse or interfere with anything provided for health and safety.

14. What do the initials PPE stand for?
personnel protective equipment.

15. Where would you use the following protective equipment?
Dust mask/respirator – when cutting materials with a disc cutter.
Safety goggles – to give eye protection when cutting materials.
Safety helmet – to give protection against falling materials or protruding objects.

16. State the type of safety equipment that you should use when handling glass.
Chrome leather gauntlets.

17. What are your responsibilities regarding safe working on site?
To act responsibly and to conduct all your activities in a safe manner.

18. List the shape and colour of the following types of safety signs.
Prohibition – circular with red border and cross bar, black symbol on a white background.
Mandatory – circular with a white symbol on a blue background.
Warning – triangular with a yellow background with a border and symbol.
Information – square or rectangular with a white symbol on a green background.

19. What do the following symbols mean?

a Caution risk of electric shock

b Caution, risk of fire

c Caution, toxic hazard

d Smoking or naked flame prohibited

e Eye protection must be worn

f Wash hands

20. Why should you not use drugs or drink alcohol when you are at work?
You could be putting yourself and others at risk.

21. What are the priorities of first aid?
save life
prevent further injuries
evacuate to medical help as soon as possible.

22. What procedures should you take if first aid is needed?
Call for help.
Send someone to telephone for an ambulance if necessary.
Do not move the casualty unless in immediate danger.
Remain with the casualty and give reassurance.
Make the casualty as comfortable as possible.
Do not allow the casualty to smoke, eat or drink.

23. What type of fire extinguisher would you use to put out the following types of fire?
Electrical – carbon dioxide or vaporized liquid.
Adhesives – foam, carbon dioxide or vaporized liquids.
Propane – water, foam, carbon dioxide or vaporized liquids.

24. Describe the procedures that you should follow if you find a fire.
Raise the alarm and call the fire brigade.
Close all doors and windows to stop the spread of fire.
Evacuate the building.
Fight the fire with an extinguisher provided but do not put yourself at risk.

25. What considerations should you take when lifting something heavy?
always use mechanical means if possible
be aware of your own capabilities
decide if it is a one-man task or do you need help
is there a clear pathway?
be sure you know the weight of the load
wear gloves to protect your hands
wear safety boots
get the feel of the load before lifting it.

26. What are the basic rules for team lifting?
everyone should be of a similar build
one person should take charge.

27. State the action to be taken to prevent fires.
do not hang clothing over or near heaters
do not let flammable rubbish accumulate
do not smoke in prohibited areas
do not overload electric sockets
switch off any electric equipment not in use.

28. Give three reasons for keeping your work area tidy.
safety
health
economy.

29. Give four building-site operations where you would insist on wearing PPE?
    Safety helmets/boots – at all times.
    Safety goggles – when carrying out any operations likely to produce chips, dust or sparks.
    Dust masks – where dust is likely to be produced.
    Gloves – when handling materials.

30. List the rules of manual handling.
    keep your back straight
    keep your arms straight
    avoid sudden movement
    grip loads with your palm of your hand, not just your fingers.

2

SETTING OUT A SMALL BUILDING

What you need to know to complete this chapter:

- how to identify/take off details from working drawings;
- how to work with scale rules;
- how to work in metric and use a tape measure;
- how to set out a rectangular building;
- how to construct profiles;
- how to transfer datums.

GLOSSARY OF TERMS

Boning rods – wooden T-pieces used in setting out drains, leveling etc.

Builders square – this is constructed from 75 mm × 50 mm timber (see below) and is used for setting out right angles.

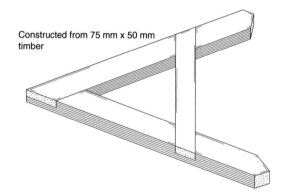

Constructed from 75 mm x 50 mm timber

Building line – an imaginary line that the face of the building must not project beyond. The Local Authority establishes it.

Cowley level – a simple form of leveling instrument consisting of a metal case containing a system of mirrors and prisms.

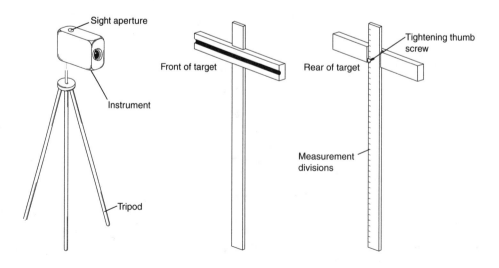

Sight aperture

Instrument

Tripod

Front of target

Rear of target

Tightening thumb screw

Measurement divisions

Datum – a fixed reference point from which all levels are taken (see temporary bench mark).

Diagonals – this is the final stage to complete the setting out by measuring the diagonals, which must be the same if the building is square. If they are not the same size then a mistake has been made in the setting out process.

Elevation – a construction drawing showing the view of a vertical surface of a building.

Hatchings – the term given to the markings on the cross-sectional drawing used to indicate the materials that it is constructed from. The official British Standards can be found in BS 1192.

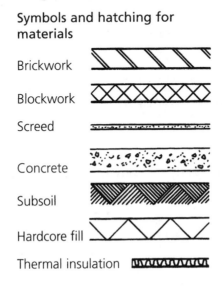

Symbols and hatching for materials

Brickwork

Blockwork

Screed

Concrete

Subsoil

Hardcore fill

Thermal insulation

Isometric – a drawing to scale, showing an oblique view of an object from a high viewing point.

Ordnance bench mark – datum which marks the height above mean sea level at Newlyn in Cornwall. These datum's are established by the Government Ordnance Survey Office and are usually permanently marked on public buildings etc.

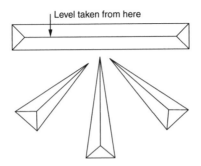

Level taken from here

Plan – a constructional drawing showing a view of a building. A floor plan shows the floor area of a building with walls in horizontal section.

Profile boards – temporary timber boards erected outside the enclosing walls of a structure at corners and used to fix string lines when setting out foundations and walls. Profiles should be constructed from 50 mm × 50 mm pegs with a 75 mm × 30 mm cross board.

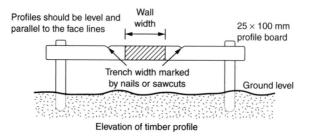

Elevation of timber profile

Ranging line – a line stretched between profiles to mark the position of a wall. Lines are made from nylon or hemp. Nylon stretches more but will retain its tension when strung between two pegs or profiles.

Scale – the proportional relationship between a representation of an object on a constructional drawing and its size, e.g. 1:10.

Section – a constructional drawing showing a view of the cut surface that would be seen it a building was cut through it.

Sight square – this is an optical instrument used for setting out right angles. It consists of an instrument head, which contains two telescopes set permanently at right angles to each other.

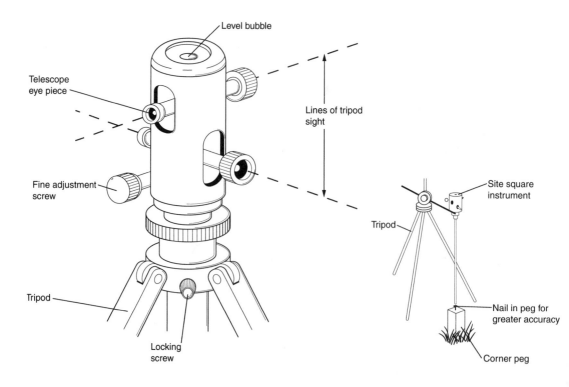

Straight edge – a purpose-made length of timber with parallel sides. It is used in conjunction with a spirit level as a method of transferring a level from one point to another.

Temporary bench mark – an assumed datum that is set up on building sites and is given a sufficient value to ensure that the lowest point on the site will still be above zero.

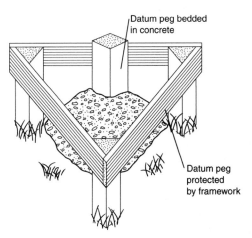

3:4:5 method – used for setting out and checking right angles. It is based on the fact that a triangle, which has sides in the ratio of 3:4:5 must contain one angle of 90°.

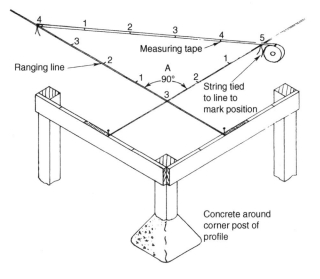

Assessments 2.1, 2.2 and 2.3

SETTING OUT A SMALL BUILDING

Time allowed

Section 2.1: ¾ hour
Section 2.2: ¾ hour
Section 2.3: 1 hour

Instructions

- You will need to have the following:
    Question paper
    Answer sheet
    Pencil.
- Ensure your name and date is at the top of the answer sheet.
- When you have decided a correct answer, draw a straight line through the appropriate letter on the answer sheet.
- If you make a mistake with your answer, change the original line by making it into a cross and then put a line through the amended answer. There is only one answer to each question.
- Do not write on the question sheet.
- Make sure you read each question carefully and try to answer all the questions in the time allowed.

Example

a	150 mm		~~a~~	150 mm
b	75 mm		b	75 mm
~~c~~	225 mm		✗	225 mm
d	300 mm		d	300 mm

Section 2.1

1. Hatching is a method which is used to indicate:
    a   materials
    b   scale
    c   components
    d   dimensions.

2. Which drawing would you use when setting out a building?
    a   site plan
    b   block plan
    c   detailed drawings
    d   sectional drawings.

3. Which of the following hatching represents brickwork?

a   c

b   d

4. The building line is fixed by the:
   a local authority
   b building inspector
   c clerk of works
   d health and safety inspector.

5. A builders square is used for:
   a setting out the trenches
   b setting up the profiles
   c setting up the corners
   d setting out the datum.

6. An assembly drawing shows:
   a The location of the building in relation to the setting out point.
   b Identifies the site and locates the outline of the building.
   c All information necessary for the manufacture and application of components.
   d In detail the construction of buildings, and junctions in and between elements etc.

7. The symbol indicated below is a:
   a sink
   b shower tray
   c wash basin
   d gulley.

8. A datum level is:
   a the ground level
   b a point from which all levels are taken
   c the foundation level
   d the floor level.

9. A site plan identifies the position of:
   a doors and windows
   b the location and the site
   c the building on the site
   d the junction between various components.

10. Components drawings specify the:
    a proposed position of the building and the site
    b proposed drainage layout
    c junction between various components
    d proposed building itself.

11. When setting out a building, what operation should be done first?
    a establish the building line
    b set out the datum pegs
    c mark out the trench line
    d set up the profile boards.

12. What is the name of the optical level, which contains a set of mirrors?
    a spirit level
    b titling level
    c quick set level
    d cowley level.

13. Which of the following hatching represents blockwork?

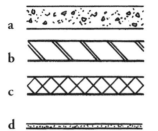

    a

    b

    c

    d

14. The baseline from which all the setting out on the site is taken is:
    a the line of the extreme edge of the foundation
    b the building line
    c the front of the furthermost projection from the main walls, e.g. porch
    d the centre line of the wall.

15. A profile when applied to setting out is:
    a a point where all levels are taken from
    b a piece of hardboard cut to the shape of a wall curved on plan
    c a framework for fixing setting out lines to
    d a piece of hardboard cut to the shape of a brick.

16. A cowley site square is effective over:
    a 30 m
    b 90 m
    c 100 m
    d 50 m.

17. The symbol indicated below is a:
    a sink
    b shower tray
    c gulley
    d manhole.

    | M.H. |

18. A 90° corner can be set out by using a tape measure by using the following ratios:
    a 2:3:4
    b 1:2:3
    c 3:4:5
    d 4:5:6.

19. The timber used to construct a profile board is made from:
    a 100 × 20 mm section
    b 100 × 50 mm section
    c 75 × 50 mm section
    d 75 × 30 mm section.

20. What is the name of the optical instrument used for setting out a 90° corner?
    a site square
    b set square
    c site datum
    d site level.

21. A detailed drawing is normally drawn to a scale of:
    a 1:50
    b 1:20
    c 1:5
    d 1:2.

22. The figure below shows a typical site square. The part marked ✕ is called a:
    a telescope eye piece
    b locking screw
    c level bubble
    d tripod.

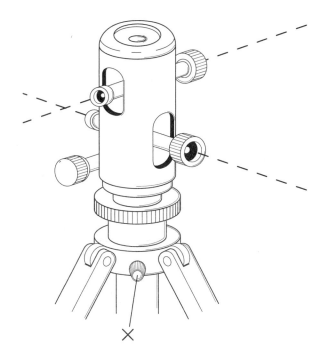

✕

23. After setting out the building you check the diagonals to make sure that the:
   a   profiles are level
   b   site is level
   c   corners are square
   d   foundations are level.

24. The working drawing shown below is called a:
   a   site plan
   b   block plan
   c   elevation
   d   section.

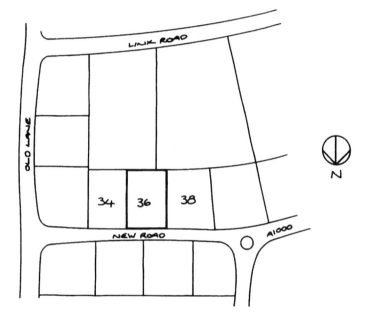

25. Why do you place the corner profiles approximately 5 m from the excavation?
   a   So that excavated material can be piled up.
   b   To leave room for the mechanical excavator.
   c   To leave room for bricks and mortar to be stacked up.
   d   To leave room for access.

Now check your answers from the grid

Q 1; a	Q 6; d	Q 11; a	Q 16; b	Q 21; d
Q 2; a	Q 7; c	Q 12; d	Q 17; d	Q 22; b
Q 3; d	Q 8; b	Q 13; c	Q 18; c	Q 23; c
Q 4; a	Q 9; c	Q 14; b	Q 19; d	Q 24; b
Q 5; c	Q 10; c	Q 15; c	Q 20; a	Q 25; b

Section 2.2

1. If a working drawing is made to a scale of 1:50, a length of 7.5 m would be shown on the drawing by a length of:
   a   55 mm
   b   150 mm
   c   275 mm
   d   550 mm.

2. The drawing shown below is called a:
   a   site plan
   b   block plan
   c   detailed drawings
   d   section drawing.

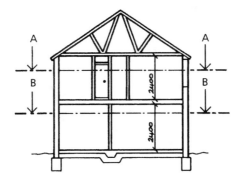

3. Corner profiles for setting out a building may be constructed from:
   a   2 pegs and 2 boards
   b   3 pegs and 3 boards
   c   3 pegs and 2 boards
   d   2 pegs and 3 boards.

4. How would you check a rectangular building for squareness?
   a   use a site square
   b   check each side to confirm the dimensions
   c   use a builders' square on all corners
   d   measure the diagonals.

5. The figure on page 32 shows a typical site square. The part marked X is called a:
   a   locking screw
   b   tripod
   c   telescope eye piece
   d   level bubble.

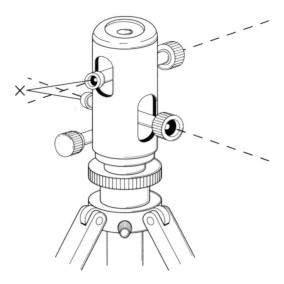

6. When excavating the foundation trench by hand, the corner profiles should be placed:
   a   1.5 m away
   b   5 m away
   c   3 m away
   d   0.5 m away.

7. The working drawing shown below is called a:
   a   site plan
   b   component drawing
   c   elevation
   d   block plan.

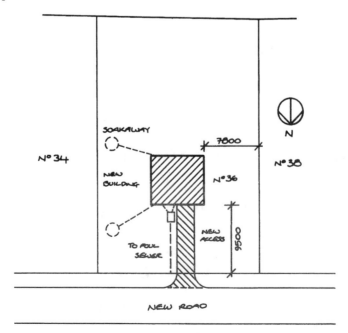

8. In the figure below, one of the diagonals is longer than the other. Which of the four pegs will need to be adjusted?

   a A

   b B

   c F

   d G.

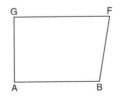

9. The cross-sectional dimension of a peg used in setting out is:

   a 50×20 mm section

   b 100×50 mm section

   c 75×50 mm section

   d 50×50 mm section.

10. What is the purpose of profiles used in setting out?

   a a means of holding the ranging line in their correct position

   b a means of holding the bricklayers line in the correct position

   c a means of marking out the shape of a bay window

   d a means of marking out a voussoir.

11. What is the name of the drawing shown below?

   a block

   b elevation

   c site

   d section.

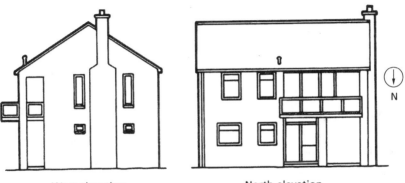

West elevation   North elevation

12. When viewing the staff through a cowley level, which of the following is showing level?

a 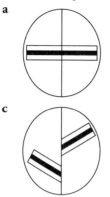   b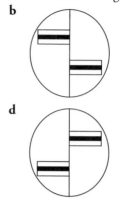

c 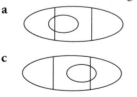   d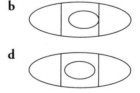

13. Boning rods are used in sets of:
- **a** 2
- **b** 3
- **c** 4
- **d** 5.

14. When setting out a right angle corner, two sides measure 6 m and 8 m. What is the length of the third and longest side?
- **a** 7.5 m
- **b** 5 m
- **c** 10 m
- **d** 12 m.

15. Which bubble is showing level?

a   b

c   d

16. What do the initials OBM stand for?
- **a** ordinary bench mark
- **b** ordnance bench mark
- **c** orderly bench mark
- **d** ordnance bend mark.

17. Which of the following hatchings represent hardcore?

a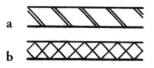

b

c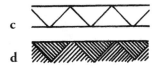

d

18. A cowley level is accurate over a distance of:
    a 30 m
    b 50 m
    c 100 m
    d 90 m.

19. What information can be found on a profile?
    a centre line of wall
    b width of trench and wall
    c width of trench only
    d width of foundation only.

20. A site plan is normally drawn to a scale of:
    a 1:50
    b 1:20
    c 1:5
    d 1:200.

21. The symbol indicated below is a:
    a gulley
    b sink
    c manhole
    d opening in a wall.

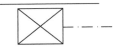

22. The hatching shown below is
    a insulation
    b planed timber
    c topsoil
    d plywood.

23. Which of the following hatchings represents concrete:

a

b

c

d

24. The symbol of a window shown is a:
    a   fixed window
    b   side-hung window
    c   top-hung window
    d   bottom-hung window.

25. When setting out a right angle corner, if one side measures 900 mm and the other side measures 1200 mm, what is the length of the diagonal?
    a   500 mm
    b   1500 mm
    c   2000 mm
    d   1000 mm.

Now check your answers from the grid

Q 1; b	Q 6; a	Q 11; b	Q 16; b	Q 21; a
Q 2; d	Q 7; a	Q 12; a	Q 17; c	Q 22; a
Q 3; c	Q 8; c	Q 13; b	Q 18; a	Q 23; c
Q 4; d	Q 9; d	Q 14; c	Q 19; b	Q 24; d
Q 5; c	Q 10; a	Q 15; d	Q 20; d	Q 25; b

Section 2.3

1. Why must you take care when setting out?

2. Why should you reverse a straight edge each time you move it?

3. Which type of tape is more accurate?

4. How should you take care of a tape?

5. What is a cowley level?

6. What is a site square?

7. What is the purpose of written specifications?

8. What information would you find on the following drawings?
   elevations
   plans
   sections
   site plans.

9. What is the purpose of using hatchings?

10. Complete the hatchings for the following:
    brickwork
    blockwork
    subsoil
    concrete
    hardcore
    insulation board.

11. Sketch the symbol for the following:
    sink
    bath
    side-hung window
    rodding eye.

12. What is the 'building line'?

13. What is the base line?

14. If a hedgerow marks the boundary, where do you take the measurement from?

15. What are profile boards and where would you find them?

16. What information can you find on a profile board and how can it be marked out?

17. Why does the method of excavation determine the position of the profile boards?

18. What is the site datum?

19. How do you protect the datum peg from being damaged?

20. What is the purpose of checking the diagonals of the building when you have set it out?

21. Why do you not use pencil lines on the profile board?

22. What are profiles constructed out of?

23. What information would you find on a title panel?

24. How do you take care of a cowley level?

Model answers for Section 2.3

1. Why must you take care when setting out?

   Mistakes during the setting-out stage can be very expensive to put right. Always check and recheck before starting any work.

2. Why should you reverse a straight edge each time you move it?

   When levelling with a straight edge and level, it should be reversed end on end each time it is used. This will counter any inaccuracies in the straight edge or level.

3. Which type of tape is more accurate?

   Steel tapes are more accurate than linen ones because the latter tend to stretch or shrink with age.

4. How should you take care of a tape?

   Clean and lightly oil at end of each day.
   Do not leave it lying around.

5. What is a cowley level?

   This is a simple form of levelling instrument and is widely used by builders for transferring levels on site.

6. What is a site square?

   This is an optical instrument used for setting out right angles on site.

7. What is the purpose of written specifications?

   The specifications contain all essential information and job requirements that will affect the price of the work but cannot be shown on the drawings.

8. What information would you find on the following drawings?

   Elevations
   External finishes to walls and roof.
   Position of windows and door openings.

   Plans
   Overall dimensions of the building.
   Position of internal walls, room sizes, doors and window openings.
   Position of fitments.

   Sections
   Provide vertical dimensions.
   Provide constructional details of foundations, floors, walls, roofs, damp proof membranes and ground levels.

   Site plans
   Gives the position of the proposed building and the general site layout of roads, services and drains.

9. What is the purpose of using hatchings?

Hatchings are used on cross-sectional drawings to indicate the materials that it is constructed of.

10. Complete the hatchings for the following:

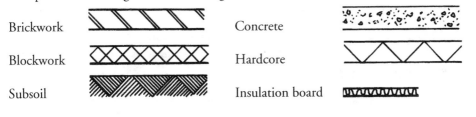

Brickwork   Concrete

Blockwork   Hardcore

Subsoil   Insulation board

11. Sketch the symbol for the following:

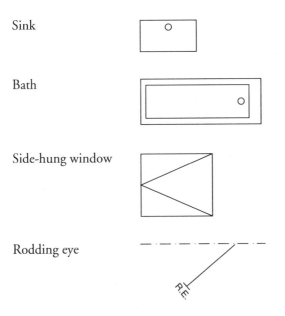

Sink

Bath

Side-hung window

Rodding eye

12. What is the 'building line'?

The local authority sets the building line and no part of the building must project beyond this line.

13. What is the base line?

The building may be placed on the building line or behind it, this is called the base line.

14. If a hedgerow marks the boundary, where do you take the measurement from?

The centre line of the hedgerow.

15. What are profile boards and where would you find them?

Profile boards provide a means of holding the ranging lines in their correct position above the ground and secure away from the point of excavation.

16. What information can you find on a profile board and how can it be marked out?

Face of the external wall
Face of the internal wall
Width of the foundation.

17. Why does the method of excavation determine the position of the profile boards?

If the foundation is to be excavated by hand then the profiles may be placed approximately 1.5 m away. If a mechanical excavator is to be used then the profiles must be placed far enough away to allow access for the excavator.

18. What is the site datum?

This is the level from which all vertical dimensions on the building will be taken from. The engineer transfers it to the site from the Ordnance Survey benchmark located nearby.

19. How do you protect the datum peg from being damaged?

The peg should be surrounded in concrete then protected by a wooden fence.

20. What is the purpose of checking the diagonals of the building when you have set it out?

By checking that the diagonal measurements are both the same, it will ensure that the building has been set out square.

21. Why do you not use pencil lines on the profile board?

Pencil lines can weather away over a period of time.

22. What are profiles constructed out of?

Pegs – 50 mm × 50 mm; cross boards – 75 mm × 30 mm.

23. What information would you find on a title panel?

Site address, clients name, drawing number, date of drawing, scales used.

24. How do you take care of a cowley level?

Always make sure the locking screw is tight before moving it, and packing it away.
Never carry the instrument on the tripod.
Always wipe clean and dry the equipment on completion of work.
The cross-target should be removed from the staff to reduce the risk of damage.

3

BRICK WALLING DETAILS

To tackle the assessments in this chapter you will need to know:

- the reasons for bonding brickwork;
- how to recognize types of brickwork bonds;
- how to set out walls;
- how to joint walls;
- how to prepare materials for use;
- how to construct half-brick and one-brick walls;
- how to recognize the different types of arches.

GLOSSARY OF TERMS

Abutments – the brickwork on either side supporting an arch.

Actual size – the size of an individual brick or block as measured on site. It may vary from the work size, within certain allowances for tolerance.

Arch – an arrangement of bricks that span an opening. It is usually curved in shape, but it may also be flat.

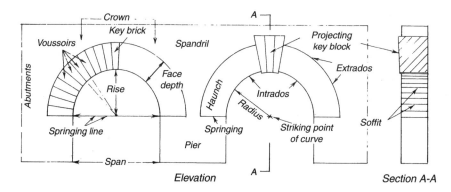

Arris – any sharp edge of a brick formed by the junction of two faces.

Axed arch – an arch formed of voussoirs that are cut to a wedge shape with a brick hammer and bolster chisel.

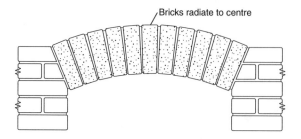

Bricks radiate to centre

Bat – a part of a brick used in bonding brickwork at corners and ends of walls, e.g. half bat or three-quarter bat.

Bed – the horizontal layer of mortar on which a brick is laid.

Bed face – the face of a brick usually laid in contact with the mortar bed.

Bed joint – the horizontal joint in the brickwork.

Bolster – a broad-bladed chisel (100 mm) made of hardened steel used for cutting bricks or blocks.

Bond – the arrangement of brick or blocks, usually overlapping to distribute the load.

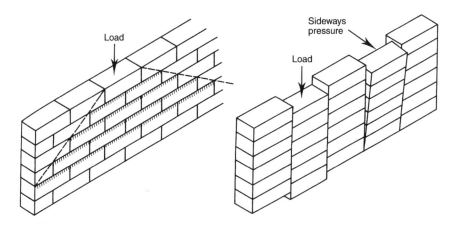

Bonding bricks – part bricks, e.g. half bats or special bricks used to bond the brickwork at features, corners and ends of walls.

Brick gauge – a wooden tool use to assist the accurate marking for cutting bricks.

Bricks – see calcium silicate, clay, common, concrete, engineering, extruded wire-cut, facing, fletton, flint lime, handmade, perforated, pressed, sand lime.

British Standards – national standards that define the sizes and properties of materials and their proper use in building.

Broken bond – occurs where the number of bricks will not fit exactly into the required length of a wall so that a cut brick has to be built into the wall.

The ideal solution is seldom possible.

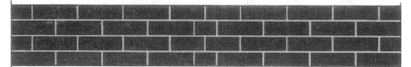

In this situation a bricklayer will usually use broken bond, located centrally.

However, some architects may prefer three-quarter bricks at each end.

Reversing the bond at each end of a wall may also be considered preferable to broken bond.

Calcium silicate brick – a brick made from sand and lime with the addition of crushed flint, autoclaved in steam under pressure.

Camber – a very flat upward curve.

Camber arch – an arch with a slight upward curvature.

Centre – a temporary frame shaped to support an arch during its construction, then removed when it has set.

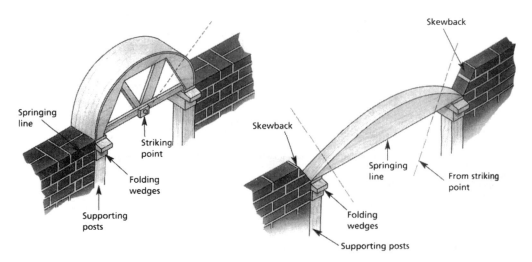

Chases – recesses cut into walls to accommodate wires or pipes.

Clay brick – a brick made from clay and fired in a kiln to produce a hard semivitreous brick.

Closers – bricks cut along the length of a brick to expose half a header (46 mm wide) in the surface of the wall and used to obtain a quarter-lap bond in walls.

Club hammer – heavy hammer used for striking a bolster chisel when cutting bricks.

Collar joint – a continuous vertical joint parallel to the face of a wall, formed in walls that are one brick or over in thickness.

Common bricks – bricks for general purpose use where the appearance does not matter.

Concrete brick – a brick made from crushed rock aggregate mixed with Portland cement.

Co-ordinate size – size of a co-ordinate space allocated to a brick or block, including the size of a mortar joint.

Coping – placed on top of a wall to provide protection from the elements. It is grooved and overhangs the wall surface to throw water clear.

Corner block – a wooden or plastic block to provide a temporary fixing at the ends of a wall for a bricklayer's line, used to keep the wall level.

Course – a row of bricks or blocks laid on a mortar bed.

Cross-joint – vertical mortar joint at right angles to the face of the wall (sometimes called a 'perp').

Crown – the highest point of the extrados.

Engineering brick – a type of clay brick traditionally used for civil engineering work where great strength and density are needed. They are defined by compliance with minimum compressive strength and maximum water absorption values.

Extrados – the external or top curve of the arch.

Extruded wire-cut bricks – bricks formed by forcing stiff moist clay, under pressure, through a die and cutting the extruded shape into individual bricks with taut wires.

Face work – brickwork or blockwork built neatly and evenly without applied finish.

Facing bricks – used in the exposed surface of brickwork where consistent and acceptable appearance is required.

Fire clay – a type of clay containing a high proportion of silica, used in the manufacture of firebricks and bedding of fire bricks in position.

Fletton bricks – semidry pressed bricks made from clay, originally made in Fletton, near Peterborough.

Flint lime bricks – see calcium silicate brick.

Frog – an indentation in one or both bed faces of some types of bricks.

Gauge rod – a batten marked at intervals for the vertical setting out of brick courses.

Plate

Head

Cill

Gauge rod

Additional information may be added to the gauge rod to remind the bricklayer of window cill and head heights or the height of the wall plate

Gauged arch – an arch built of purpose-made or carefully cut wedge-shaped bricks jointed with a non-tapered mortar joint.

Handmade bricks – bricks molded to shape by hand from moist clay.

Haunch – the lower third of the arch.

Header – the end face of a standard brick.

Hod – a three-sided box mounted on a pole handle, used over the shoulder for carrying small quantities of bricks or mortar.

Indent – a recess formed in the brickwork ready to receive a new wall at a later date.

Intrados – the underside or soffit line of an arch.

Joint profile – the shape of the mortar joint finish.

Jointing – forming the finished surface profile of a mortar joint as the work proceeds, without pointing.

Key brick – the central brick at the crown of an arch.

Lap – the amount of horizontal distance from one perpend on one course to the perpend other on the course above and below. This is normally half lap on half brick walls and quarter lap on one-brick walls or over.

Level (1) – the horizontality of courses of brickwork.

Level (2) – see spirit level.

Line (1) – a string line used to guide the setting of bricks to line and level.

Line (2) – the straightness of the brickwork.

Lintel – a horizontal member spanning an opening to support the structure above. May be made from wood, steel or concrete.

Perforated bricks – extruded wire-cut bricks with holes through from bed face to bed face.

Perpends (perps) – the vertical lines controlling the verticality of cross-joints appearing in the face of a wall.

Pier – the thickening of a wall to improve its strength. Can be attached or detached from the wall, i.e. gate pier.

Plugging chisel – a stout chisel with a narrow cutting edge for cutting out hardened mortar from a joint between bricks.

Plumb – the verticality of brickwork.

Pointing – finishing mortar joints by raking out part of the jointing mortar, filling with additional mortar, and working it to form the finished profile.

Polychromatic brickwork – decorative patterned work, which features bricks of different colours.

Pressed bricks – bricks formed by pressing moist clay into shape by a hydraulic press.

Quoin – the external corner of a wall.

Racking back – temporarily finishing each brickwork course in its length short of the course below so as to produce a stepped diagonal line to be joined with later work.

Reference panels – a panel of brickwork built at the start of a contract to set standards of appearance and workmanship.

Repointing – the raking out of old mortar joints and replacing it with new.

Return corners – the portion of wall at right angles to the main wall.

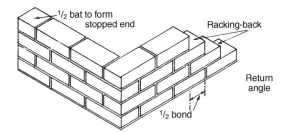

Reverse bonds – the placing of half bats at alternative ends on each course to avoid broken bond in the work.

Rise – the vertical distance between the springing line and the highest point on the intrados.

Rough arch – an arch formed with bricks not cut to shape and with tapered joints.

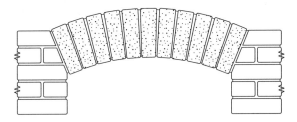

Sample panel – a panel of brickwork, which may be built to compare materials and workmanship with those of a reference panel.

Sand lime bricks – see calcium silicate brick.

Scutch – a hammer with sharp-edged teeth or comb blade. Used for trimming a cut brick or block to shape.

Size – see actual size, co-ordinate size and work size.

Skewback – an angled cut brick, which provides an inclined surface from which an arch springs.

Soffit – the under side of an arch, lintel or overhang of a roof.

Span – the distance between the abutments.

Spirit level – a device used for checking the horizontally or verticality of brickwork etc. It consists of one or more sealed glass tubes, each containing liquid and an air bubble, mounted in a frame.

Spot board – a board up to 1 m² on which fresh mortar is placed ready for use.

Springing line – the horizontal line between the springings of an arch.

Springing plane – at the end of an arch, which springs from a skewback.

Stop end – a three-sided box shaped shoe of DPC material sealed to the end of a DPC tray to divert the discharge of water.

Stopped end – the termination of a wall, which should be correctly bonded to coincide with the bond used for the wall.

Storey rod – gauge rod of a storey height with additional marks to indicate features such as lintel, sill floor joists etc.

Stretcher – the longer face of a brick showing in the surface of a wall.

Striking point – the centre of the circle of which the intrados and extrados are arcs.

Struck – pointing of the mortar by pressing the top inwards with the edge of the trowel.

Tingle plate – a metal plate shaped to give intermediate support to a line when building long lengths of walls.

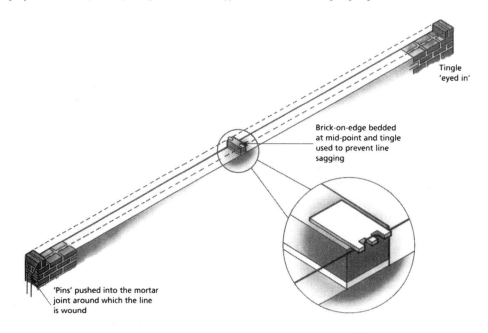

Tingle
'eyed in'

Brick-on-edge bedded
at mid-point and tingle
used to prevent line
sagging

'Pins' pushed into the mortar
joint around which the line
is wound

Tolerance – allowable variation between a specified dimension and an actual dimension.

Toothing – leaving the vertical end of a wall unfinished in its bond to enable the wall to be continued at a later stage.

Traverse joint – a joint that passes from the face of the wall to the rear of the wall. It normally occurs in walls that are 1½ bricks thick.

Turning piece – a centre cut out from one piece of timber.

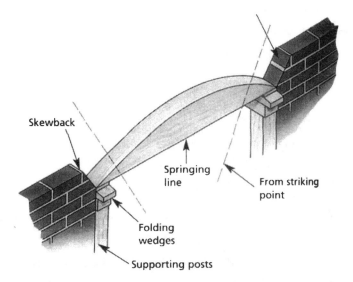

Skewback

Springing
line

From striking
point

Folding
wedges

Supporting posts

Voussoir – a wedge-shaped brick used in a gauged arch.

Work size – the size of a brick or block specified for its manufacture. It is derived from the co-ordinate size less the allowance for mortar joints.

Assessments 3.1, 3.2 and 3.3

BRICK WALLING DETAILS

Time allowed

Section 3.1: 1 hour
Section 3.2: 1 hour
Section 3.3: 1½ hours

Instructions

- You will need to have the following:
  Question paper
  Answer sheet
  Pencil.
- Ensure your name and date is at the top of the answer sheet.
- When you have decided a correct answer, draw a straight line through the appropriate letter on the answer sheet.
- If you make a mistake with your answer, change the original line by making it into a cross and then put a line through the amended answer. There is only one answer to each question.
- Do not write on the question sheet.
- Make sure you read each question carefully and try to answer all the questions in the time allowed.

Example

a	150 mm	a̶	150 mm
b	75 mm	b	75 mm
c̶	225 mm	✗	225 mm
d	300 mm	d	300 mm

Section 3.1

1. The sketch below shows an example of which type of joint finish?
   a   tooled joint
   b   flush joint
   c   recess joint
   d   weather struck joint.

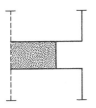

2. A brick scutch should be used to:
   a adjust the quoin brick
   b cut small bevelled bats
   c trim rough cut bricks
   d cut perforated bricks.

3. English garden wall bond consists of:
   a alternate courses of headers and stretchers
   b three courses of headers to one course of stretchers
   c three stretchers to one header in each course
   d three courses of stretchers to one course of headers.

4. In English and Flemish bond, the quoin header is normally followed by a:
   a header
   b stretcher
   c half bat
   d closer.

5. When broken bond occurs in straight lengths of walls, the smallest brick cut is a:
   a three-quarter brick
   b half brick
   c quarter brick
   d full brick.

6. The horizontal distance by which one brick overlaps the brick below is called:
   a bond
   b lap
   c half bond
   d quarter bond.

7. In Flemish garden wall bond, the number of stretchers following each header is:
   a three
   b two
   c four
   d one.

8. A rough arch consists of:
   a uncut bricks with wedge-shaped joints
   b cut bricks with wedge-shaped joints
   c uncut bricks with parallel joints
   d cut bricks with parallel joints.

9. To keep the bed joints to the required thickness it is advisable to:
   a level the work frequently
   b lay the brick frog up
   c use a gauge lath/rod
   d lay the brick frog down.

10. A half-brick wall is 7 m × 2 m high. What is the number of bricks required to build the wall?
  a 960
  b 840
  c 920
  d 880.

11. The area of a brick which surrounds the frog is called:
  a margin
  b stretcher face
  c arris
  d header face.

12. The term 'racking back' is used to describe:
  a raking out the joints of brickwork ready for pointing
  b stepped effect obtained when setting up corners
  c slope of walls built at less than 90°
  d sloping sides of a trench.

13. One course of Flemish bond consists of:
  a alternate courses of headers and stretchers
  b all courses headers
  c all courses stretchers
  d alternate headers and stretchers in the same course.

14. When easing an arch centre it is advisable to:
  a remove the nails
  b take out the props
  c tap it gently with a hammer
  d loosen the wedges.

15. What is sectional bond?
  a the bond across the wall
  b the joint parallel to the wall face
  c correct bonding of the wall face
  d bonding of the internal angle.

16. The main reason for using a garden wall bond is to keep a fair face on both sides of a wall by:
  a reducing the number of closers
  b reducing the number of headers
  c increasing the number of headers
  d increasing the number of closers.

17. The standard gauge for four courses of bricks is:
  a 300 mm
  b 450 mm
  c 750 mm
  d 675 mm.

18. The purpose of a turning piece is to:
    a   set out circular work
    b   assist in building a bullseye
    c   provide a template for curved work
    d   provide temporary support for arches.

19. The joint finish below is called a:
    a   flush joint
    b   recess joint
    c   struck joint
    d   weather struck joint.

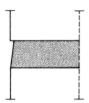

20. A temporary brick pier bedded to gauge and plumbed is called a:
    a   dead man
    b   isolated pier
    c   profiles
    d   attached pier.

21. When using a recessed joint how deep should the joint be raked out?
    a   15 mm
    b   10 mm
    c   20 mm
    d   12 mm.

22. The sloping brick supporting a segmental arch is called the:
    a   springer
    b   skewback
    c   striking point
    d   haunch.

23. The cut brick shown below is called a:
    a   queen closer
    b   king closer
    c   bevelled closer
    d   prince closer.

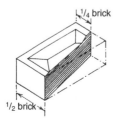

24. The shaded brick indicated with an × in the figure below is called a:
   a header
   b stretcher
   c tie
   d closer.

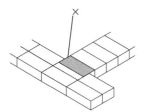

25. The usual thickness for a joint between the voussoirs in a gauged arch is:
   a 3 mm
   b 10 mm
   c 5 mm
   d 15 mm.

26. The voussoirs for an axed segmental arch are set out on the:
   a centre line
   b extrados
   c springing line
   d intrados.

27. The intrados of an arch is:
   a the inner edge
   b the outer edge
   c the highest point
   d the centre of the arch.

28. Which of the following arches is called a 'bonded' arch?

a
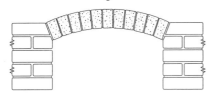

b
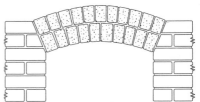

c
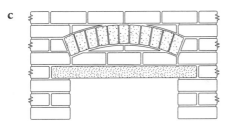

d
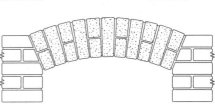

29. The part of a brick indicated by × is called the:
   a   arris
   b   frog
   c   header face
   d   stretcher face.

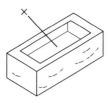

30. The appropriate angle of distribution of a load on a well bonded wall should be:
   a   65°
   b   45°
   c   22°
   d   90°.

Now check your answers from the grid

Q 1; c	Q 7; a	Q 13; d	Q 19; d	Q 25; a
Q 2; c	Q 8; a	Q 14; d	Q 20; a	Q 26; b
Q 3; b	Q 9; c	Q 15; a	Q 21; a	Q 27; a
Q 4; d	Q 10; b	Q 16; b	Q 22; b	Q 28; d
Q 5; c	Q 11; a	Q 17; a	Q 23; c	Q 29; b
Q 6; b	Q 12; b	Q 18; d	Q 24; c	Q 30; b

Section 3.2

1. The maximum depth of a vertical chase formed in the face of a wall is:
   a   a half-brick thickness
   b   a third of the thickness
   c   a quarter of the thickness
   d   an eighth of the thickness.

2. The most suitable joint finish for an internal block wall is:
   a   flush
   b   weather struck
   c   keyed
   d   recessed.

3. A brick hammer should be used to:
   a adjust the quoin brick
   b cut small bevelled bricks
   c cut perforated bricks
   d cut rough bricks.

4. Stabilizing a bricklayers line in windy conditions is done by using:
   a line pins
   b tingle plate
   c heavy lines
   d corner blocks.

5. The wall shown below is built in:
   a Flemish garden wall bond
   b English bond
   c Flemish bond
   d English garden wall bond.

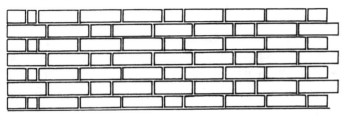

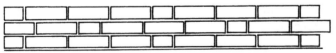

Alternative method of returning corners

6. Rough ringed arches have wedge shaped joints because:
   a the arch strength is increased
   b they improve the appearance
   c the bricks are not cut to shape
   d less mortar is used.

7. Plasticizer used in mortar increases the:
   a strength
   b density
   c adhesion
   d workability.

8. Sand used for pointing should be the same quality and batch at all times in order to maintain:
   a weathering properties
   b workability
   c colour
   d cohesion.

9. Brickwork with hand-made bricks is best finished with a flush joint because:
   a   the rough texture of the brick is emphasized
   b   it is more weather resistant
   c   flush jointing is easier to do
   d   it emphasizes the joint lines.

10. An arch centre is:
    a   the point from where the curve is taken
    b   the centre line of the arch
    c   the temporary wooden support for the arch
    d   the centre brick of the arch.

11. In bonded arches, the number of voussoirs each side of the keyed brick should be:
    a   odd in number
    b   even in number
    c   it does not matter
    d   odd on one side, even on the other.

12. A template is a:
    a   piece of plywood cut to the shape of the required voussoirs
    b   solid piece of wood cut to the shape of the arch
    c   timber centre on which the arch is built
    d   piece of plywood cut to the shape of the arch.

13. A skewback is:
    a   a type of fixing for a timber frame
    b   the slope of the weathering of a cill
    c   the slope of the brick cut to receive the arch
    d   the centre brick of the arch.

14. Toothing is normally used where walls are to be:
    a   built higher
    b   terminated
    c   corbelled out
    d   extended later.

15. When a block indent is formed to receive a 'T' junction, the number of courses per indent is:
    a   two
    b   three
    c   four
    d   five.

16. Broken bond occurs in walls when:
    a   the length of the wall does not work out in full bricks
    b   one course starts with a header and the other with a stretcher
    c   English bond is combined with Flemish bond
    d   diagonal cracking occurs through the perpends.

17. The number of bricks needed to build 1 m² of brickwork in English bond is:

   a 60

   b 90

   c 150

   d 120.

18. The figure below illustrates which type of arch?

   a relieving arch

   b bonded arch

   c double ringed arch

   d gauged arch.

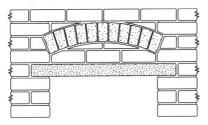

19. The underside of a segmental arch is called the:

   a skewback

   b intrados

   c haunch

   d soffit.

20. The number of bricks needed per meter square in stretcher bond is:

   a 120

   b 60

   c 90

   d 150.

21. Using the standard gauge for metric bricks, how many courses would occur in a height of 1.275 m?

   a 12

   b 15

   c 17

   d 18.

22. The term 'perpend' means:

   a the horizontal joint appearing in the wall face

   b the joint that runs through the thickness of the wall

   c the vertical joint that appears in the wall face

   d the joint that passes from face to face of the wall.

23. What does the term 'quoin' mean?

   a the top of the wall

   b the back of the wall

   c the face of the wall

   d the corner of a wall.

24. The standard gauge for six courses of bricks is:
    a 600 mm
    b 450 mm
    c 300 mm
    d 150 mm.

25. What is the term indicated by × on the figure below?
    a arris
    b lap
    c perpend
    d header.

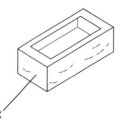

26. Vertical reinforcement within a wall is used to:
    a combat compressive loads
    b resist lateral pressure
    c prevent moisture penetration
    d strengthen weak mortar.

27. The part of a brick indicated by × is called the:
    a arris
    b frog
    c header face
    d stretcher face.

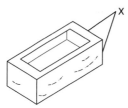

28. When carrying out BS 3921 test for brick dimensions, the number of bricks used is:
    a 24
    b 14
    c 26
    d 34.

29. The minimum gap in a cavity wall should be:
    a 10 mm
    b 25 mm
    c 50 mm
    d 75 mm.

30. The wall shown below is built in:
   a English bond
   b English garden wall bond
   c Flemish garden wall bond
   d Dutch bond.

(Note the half-bonding of the stretcher courses)

Now check your answers from the grid

Q 1; b	Q 7; d	Q 13; c	Q 19; d	Q 25; d
Q 2; a	Q 8; c	Q 14; d	Q 20; b	Q 26; b
Q 3; d	Q 9; a	Q 15; b	Q 21; c	Q 27; a
Q 4; b	Q 10; c	Q 16; a	Q 22; c	Q 28; a
Q 5; a	Q 11; b	Q 17; d	Q 23; d	Q 29; c
Q 6; c	Q 12; a	Q 18; a	Q 24; b	Q 30; b

Section 3.3

1. What is the main function of a wall?

2. Make a neat sketch of a bonded wall showing how the load is spread.

3. Why is it important to keep perpends plumb?

4. Label the brick terms on the drawing below

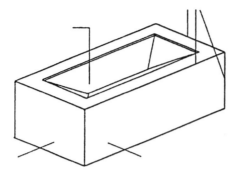

5. What does 'dry bonding' mean?

6. What is the distance that you should stack bricks etc. away from the face of the wall?

7. Why should you reverse a straight edge when transferring levels?

8. Describe how you would check a right angle corner by using the 3:4:5 method.

9. What does the term 'gauge' mean?

10. What is the purpose of jointing/pointing?

11. With the aid of a neat sketch, describe four different types of joint finishes.

12. What is the purpose of a tingle plate?

13. Why should you avoid building large corners?

14. What is broken bond?

15. Why should you avoid vertical lines when toothing?

16. What is a turning piece?

17. What is the purpose of using timber wedges?

18. Explain how you would mark out the skewback for a rough arch.

19. Explain the following terms:
    span
    springing point
    intrados
    extrados
    voussoir.

20. Describe the different classes of arches:
    rough
    axed
    relieving
    bonded.

21. What is the name of the bond that consists of three stretchers to one header in the same course?

22. Describe the following terms:
    jointing
    pointing.

23. Make a neat sketch of the correct method of making a recessed joint.

24. What is the purpose of a garden wall bond?

25. What type of brick is best for a brick coping?

Model answers for Section 3.3

1. What is the main function of a wall?

    The main function of a wall is to carry loads down to the foundation.

2. Make a neat sketch of a bonded wall showing how the load is spread.

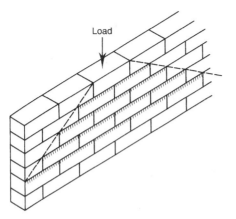

3. Why is it important to keep perpends plumb?

    It is important to keep the perpends plumb to ensure bond is maintained throughout the length of the wall.

4. Label the brick terms on the drawing below.

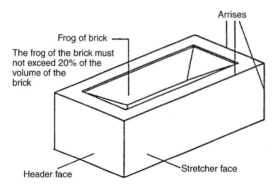

5. What does 'dry bonding' mean?

This is the term given to laying the bricks without applying a bed or cross-joint. This is done to establish the bond and size of cross-joints required for a given dimension.

6. What is the distance that you should stack bricks etc. away from the face of the wall?

Stack should be approximately 600 mm from the face of the wall to give adequate working space.

7. Why should you reverse a straight edge when transferring levels?

The straight edge should be reversed end to end to even out any inaccuracies there may be in the level or the straight edge.

8. Describe how you would check a right angle corner by using the 3:4:5 method.

This is achieved by using a tape measure. You measure 3 units along one leg, 4 units along the other leg and the diagonal should be 5 units long. This is based on Pythagorus's theorem.

9. What does the term 'gauge' mean?

Gauge is the name given to the combined depth of a brick plus the bed joint.

10. What is the purpose of jointing/pointing?

The main purpose of jointing or pointing is to seal the surface of the mortar joint to prevent the penetration of moisture, which could lead to dampness on the internal face of the building.

11. With the aid of a neat sketch, describe four different types of joint finishes.

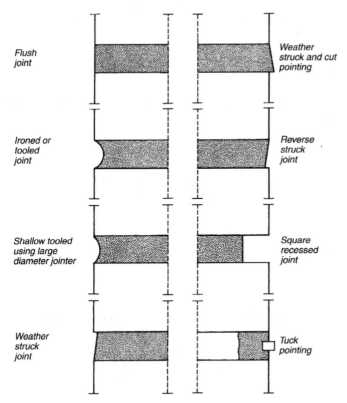

Flush joint

Ironed or tooled joint

Shallow tooled using large diameter jointer

Weather struck joint

Weather struck and cut pointing

Reverse struck joint

Square recessed joint

Tuck pointing

12. What is the purpose of a tingle plate?

On long walls it is necessary to provide intermediate support for the line to stop it sagging or swaying in windy conditions.

13. Why should you avoid building large corners?

You should avoid building large corners because it is more economical to lay to a line than use a level.

14. What is broken bond?

When a length of wall does not convert into an equal number of full bricks, cuts may have to be introduced to make up the difference by introducing cut bricks. This is termed broken bond.

15. Why should you avoid vertical lines when toothing?

Vertical lines in toothing should be avoided, as this will reduce the stability of the wall.

16. What is a turning piece?

A turning piece is used to support a segmental arch. It is normally cut out of a solid piece of timber or can be made from plywood and packed out to the width of the arch.

17. What is the purpose of using timber wedges?

    Timber wedges are used to assist in the removal of temporary timbering once the arch is built and mortar joints have hardened.

18. Explain how you would mark out the skewback for a rough arch.

    For rough arches the skewback can be established by cutting the angle from the brick placed squarely on the turning piece.

19. Explain the following terms:

    Span – the distance between the reveals of the opening of the arch span.
    Springing point – the point at which the arch intersects the two springing points.
    Intrados – the inside edge of an arch.
    Extrados – the outside edge of an arch.
    Voussoir – any wedge-shaped brick in the arch.

20. Describe the different classes of arches:

    Rough – built using full bricks and wedged-shaped joints. Used where appearance is not important.
    Axed – build using wedge-shaped bricks and parallel joints.
    Relieving – used to relieve the weight from a timber lintel.
    Bonded – the arch is bonded on its face.

21. What is the name of the bond that consists of three stretchers to one header in the same course?

    Flemish garden wall bond.

22. Describe the following terms:

    Jointing – the term used for the treatment of mortar joints at the time the bricks are laid, using the bedded mortar to form the joint.
    Pointing – the term used for the treatment of mortar joints that were recessed when the wall was built, to be filled and treated at a later date.

23. Make a neat sketch of the correct method of making a recessed joint.

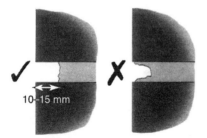

24. What is the purpose of a garden wall bond?

Garden wall bonds are used to allow a fair face to be achieved on both sides of a one-brick wall. Reducing the number of headers used in the wall does this.

25. What type of brick is best for a brick coping?

Bricks should be of Class A or B engineering quality, bedded on a strong sand–cement mix.

WALLING DETAILS

To tackle the assessments in this chapter you will need to know:

- how to build solid and cavity walls;
- how to bed to a rake;
- how to construct chimney breasts, flues and stacks;
- how to joint walls;
- how to identify/take off materials/components;
- how to prepare materials for use.

GLOSSARY OF TERMS

Actual size – the size of an individual brick or block as measured on site. It may vary from the work size within certain allowances for tolerance.

Air brick – a perforated building block, which allows air to pass through the wall.

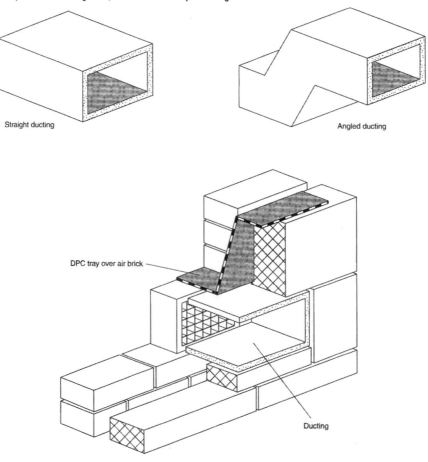

Straight ducting

Angled ducting

DPC tray over air brick

Ducting

'Cut away' section through air brick and ducting

Bearing – the amount by which a lintel or beam rests on its support.

Bond – the arrangement of brick or blocks, usually overlapping to distribute the load.

Bonding bricks – part bricks, for example half bats or special bricks used to bond the brickwork at features, corners and ends of walls.

Brick gauge – a wooden tool use to assist the accurate marking for cutting bricks.

Bricks – see calcium silicate, clay, common, concrete, engineering, extruded wire cut, facings, flettons, flint lime, hand-made, perforated, pressed, sand lime.

British Standards – national standards defining the sizes and properties of materials and their proper use in building.

Broken bond – occurs where the number of bricks will not fit exactly into the required length of a wall so that a cut brick no smaller than a half-bat is built into the wall.

Calcium silicate brick – a brick made from sand and lime with the addition of crushed flint, autoclaved under pressure.

Cavity batten – a timber batten, with lifting wires, lies across the space of the cavity to prevent mortar dropping into cavity.

Cavity tray – see DPC tray.

Cavity walling – wall built of two separate leaves tied together with wall ties with a space between them, usually at least 50 mm.

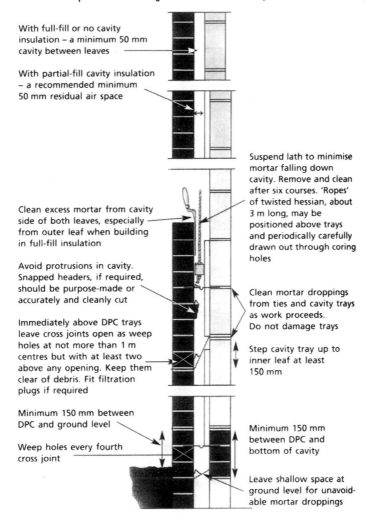

With full-fill or no cavity insulation – a minimum 50 mm cavity between leaves

With partial-fill cavity insulation – a recommended minimum 50 mm residual air space

Clean excess mortar from cavity side of both leaves, especially from outer leaf when building in full-fill insulation

Avoid protrusions in cavity. Snapped headers, if required, should be purpose-made or accurately and cleanly cut

Immediately above DPC trays leave cross joints open as weep holes at not more than 1 m centres but with at least two above any opening. Keep them clear of debris. Fit filtration plugs if required

Minimum 150 mm between DPC and ground level

Weep holes every fourth cross joint

Suspend lath to minimise mortar falling down cavity. Remove and clean after six courses. 'Ropes' of twisted hessian, about 3 m long, may be positioned above trays and periodically carefully drawn out through coring holes

Clean mortar droppings from ties and cavity trays as work proceeds. Do not damage trays

Step cavity tray up to inner leaf at least 150 mm

Minimum 150 mm between DPC and bottom of cavity

Leave shallow space at ground level for unavoid-able mortar droppings

Cellular block – concrete block with large voids (holes) that do not pass right through the block.

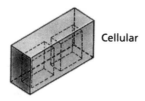

Cellular

Chases – recesses cut into walls to accommodate wires or pipes.

Chimney back – the back of the fireplace.

Chimney breast – a projecting portion of an internal wall face that contains the fireplace and the flue.

Chimney stack – the portion of the chimney construction containing the tops of the flue that passes through the roof.

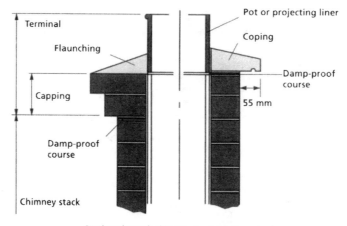

Section through chimney stack and terminal
showing alternative constructions

Chimney throating – the portion of the flue just above the lintel over the fireplace.

Cill – see sill.

Clay brick – a brick made from clay and fired in a kiln to produce a hard semivitreous brick.

Common brick – a brick for general purpose use where the appearance does not matter.

Concrete brick – a brick made from crushed rock aggregate mixed with Portland cement.

Co-ordinate size – size of a co-ordinate space allocated to a brick or block, including the size of a mortar joint.

Course – a row of bricks or blocks laid on a mortar bed.

Cross-joint – vertical mortar joint at right angle to the face of the wall (sometimes called a 'perp').

Damp proof course (DPC) – a layer or strip of impervious material placed in a joint of a wall to prevent the passage of water.

DPC membrane (DPM) – a layer or sheet of impervious material within or below a floor to prevent the passage of moisture.

DPC tray – a wide DPC bedded in the outer leaf of a cavity wall, stepping out to the inner leaf. It diverts water in the cavity through weepholes in the outer leaf.

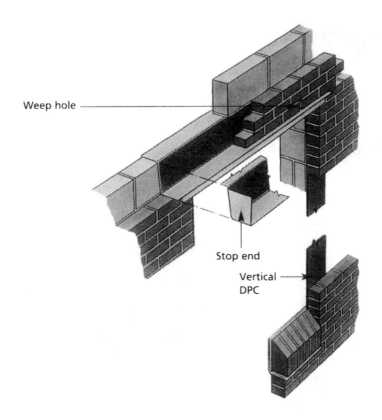

Weep hole

Stop end

Vertical DPC

Eaves – lower edge of a pitched roof or edge of a flat roof.

Engineering brick – a type of clay brick traditionally used for civil engineering work where great strength and density are needed. They are defined by compliance with minimum compressive strength and maximum water absorption values.

Extruded wire-cut bricks – bricks formed by forcing stiff moist clay, under pressure, through a die and cutting the extruded shape into individual bricks with taut wires.

Face work – brickwork or block work built neatly and evenly without applied finish.

Facing bricks – used in the exposed surface of brickwork where consistent and acceptable appearance is required.

Flashing – a water-proof sheet material, usually lead, dressed around the chimney to prevent entry of rainwater around a chimney stack and roof.

Flaunching – the cement fillet at the junction around the chimney pot and stack.

Fletton brick – semidry pressed bricks made from clay, originally made in Fletton, near Peterborough.

Flint lime bricks – see calcium silicate brick.

Flue – a pipe formed for conveying smoke or gasses from a fire.

Flue lining – precast hollow fireclay blocks, round or square in section with rebated ends, which are built into the flue during construction.

Weak lime mortar or
insulating concrete

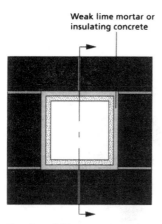

Plan view of chimney stack

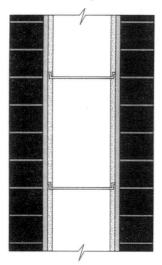

Section through lined flue

Foundation – substructure to bear on supporting sub soil. Usually made from *in situ* concrete.

Frog – an indentation in one or both bed faces of some types of bricks.

Frost damage – the destructive action of freezing water and thawing ice in saturated materials.

Gable – the portion of a wall above eave levels that forms a triangle at the sides of a pitched roof.

Gauge rod – a batten marked at intervals for the vertical setting out of brick courses.

Hand-made bricks – bricks molded to shape by hand from moist clay.

Hard-core – broken brick, stone or aggregate used for a subbase below a concrete floor or roads.

Head – the horizontal member of a door or window frame.

Hearth – the slab projecting in front of the fireplace jamb.

Insulation bat – rectangular units of resilient fibrous insulation material used to partially or fullly fill the air space in a cavity wall.

Insulation board – rectangular unit or rigid insulation material used to partially fill the air space in a cavity wall.

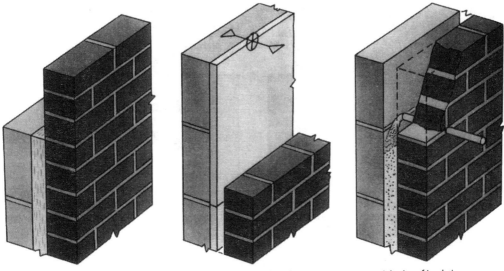

| Full-fill cavity bats | Partial-fill cavity boards | Injection of insulation |

Insulation material – material used to reduce the passage of heat through a wall.

Jamb – the brickwork on either side of an opening.

Joint profile – the shape of the mortar joint finish.

Jointing – forming the finished surface profile of a mortar joint as the work proceeds, without pointing.

Leaf – one of two parallel walls that are tied together as a cavity wall.

Lintel – a horizontal member spanning an opening to support the structure above. May be made from wood, steel or concrete.

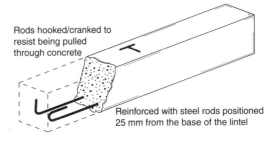

Rods hooked/cranked to resist being pulled through concrete

Reinforced with steel rods positioned 25 mm from the base of the lintel

Used to bridge openings in internal walls

Parapet wall – upper part of a wall that bounds a roof, balcony, terrace or bridge.

Partition wall – a wall within a building to divide the space within. It may or may not support floors or roofs.

Perforated bricks – extruded wire-cut bricks with holes through from bed face to bed face.

Plumb – the verticality of brickwork.

Pointing – finishing mortar joints by raking out part of the jointing mortar, filling with additional mortar, and working it to form the finished profile.

Polychromatic brickwork – decorative patterned work, which features bricks of different colours.

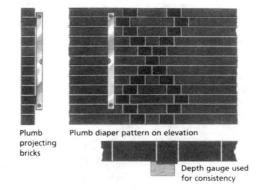

Plumb projecting bricks

Plumb diaper pattern on elevation

Depth gauge used for consistency

Pressed bricks – bricks formed by pressing moist clay into shape by a hydraulic press.

Quoin block – an L-shaped concrete block on a plan for maintaining bond at corners.

Purpose-made quoin return block

Reference panel – a panel of brickwork built at the start of a contract to set standards of appearance and workmanship.

Sample panel – a panel of brickwork, which may be built to compare materials and workmanship with those of a reference panel.

Sand lime bricks – see calcium silicate brick.

Sill – the lower horizontal edge of an opening.

Size – see actual size, co-ordinate size and work size.

Soldier course – a brick laid vertically on end with its stretcher face showing on the surface of the wall.

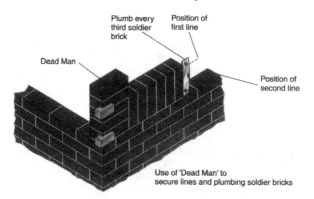

Plumb every third soldier brick

Position of first line

Dead Man

Position of second line

Use of 'Dead Man' to secure lines and plumbing soldier bricks

String course – a long narrow course projecting from the general face of the brickwork.

Storey rod – gauge rod of a storey height with additional marks to indicate features such as lintel, sill floor joists etc.

Ties – see wall ties.

Tingle plate – a metal plate shaped to give intermediate support to a line when building long lengths of walls.

Toothing – leaving the vertical end of a wall unfinished in its bond to enable the wall to be continued at a later stage.

Wall plate – a timber bedded on top of a wall for supporting roof joists.

Wall ties – a component-made non-metallic or galvanized metal or plastic, either built into the two leaves of a cavity wall to link them, or used as a restraint fixing to tie back cladding to a wall.

Types of ties

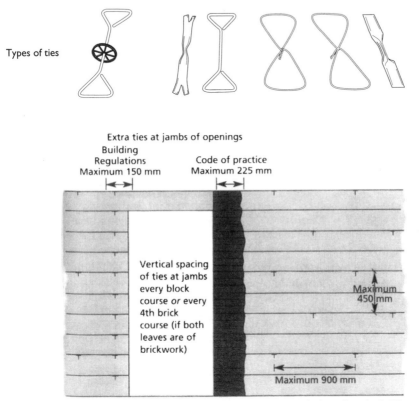

NOTE: If one leaf is less than 90 mm thick the maximum horizontal spacing is 450 mm.

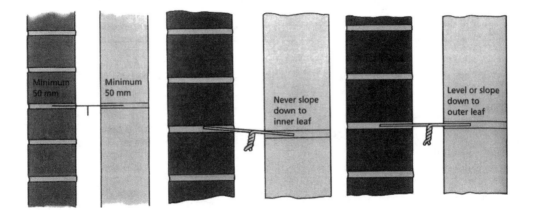

Weep hole – a hole through the brickwork, usually an unmortared cross-joint, through which water can drain to its outer face.

Work size – the size of a brick or block specified for its manufacture. It is derived from the co-ordinating size less the allowance for mortar joints.

Assessments 4.1, 4.2 and 4.3

BRICK WALLING DETAILS

Time allowed

Section 4.1: 1 hour
Section 4.2: 1 hour
Section 4.3: 1½ hours

Instructions

- You will need to have the following:
     Question paper
     Answer sheet
     Pencil.
- Ensure your name and date is at the top of the answer sheet.
- When you have decided a correct answer, draw a straight line through the appropriate letter on the answer sheet.
- If you make a mistake with your answer, change the original line by making it into a cross and then put a line through the amended answer. There is only one answer to each question.
- Do not write on the question sheet.
- Make sure you read each question carefully and try to answer all the questions in the time allowed.

Example

a	150 mm	a̶	150 mm
b	75 mm	b	75 mm
c̶	225 mm	✸	225 mm
d	300 mm	d	300 mm

Section 4.1

1. What is the spacing of wall ties at openings?
     a   150 mm
     b   450 mm
     c   75 mm
     d   300 mm.

2. Leaning chimney stacks often indicate:
   a   sulphate attack
   b   poor workmanship
   c   defective pointing
   d   roof subsidence.

3. What is the minimum width of the clear space in a partial filled cavity wall?
   a   15 mm
   b   50 mm
   c   75 mm
   d   25 mm.

4. A chimney stack measures 900 mm × 450 mm on plan; its maximum height should be:
   a   900 mm
   b   1.25 m
   c   2.0225 m
   d   2.25 m

5. How far apart should wall-plate anchor ties be placed?
   a   1.5 m
   b   1.2 m
   c   1.0 m
   d   0.5 m

6. Flue liners should be surrounded by:
   a   weak mortar mix
   b   sand/lime mortar
   c   high-alumina cement
   d   sulphate-resisting cement mortar.

7. Where would you find a 'tiled knee' course?
   a   over a window or door opening
   b   decorative sill under a window
   c   decorative corbel on a gable end
   d   under a door opening.

8. The minimum thickness of a constructional hearth is:
   a   100 mm
   b   150 mm
   c   200 mm
   d   125 mm.

9. A fender wall is built to support:
   a   jambs and constructional hearth
   b   floor joists and constructional hearth
   c   oversite concrete and floor joists
   d   floor joists and jambs.

10. Wooden fixings are normally built into door and window jambs. These are called:
    a joggles
    b pallets
    c wedges
    d plugs.

11. The material used to form the joint between the tiled surround and the fireback is:
    a cement mortar
    b tarred rope
    c lime mortar
    d fibreglass rope.

12. To comply with current building regulations, the minimum widths of a cavity wall is:
    a 50 mm
    b 60 mm
    c 75 mm
    d 100 mm.

13. A chimney stack passing through the ridge of a roof shall project a minimal amount of:
    a 600 mm
    b 450 mm
    c 750 mm
    d 900 mm.

14. Bridging of wall ties with mortar can cause:
    a vermin infestation
    b light damp patches on the outer leaf
    c damp patches on the inner leaf
    d frost damage.

15. The weathering to the top of a chimney stack is the:
    a flashing
    b flaunching
    c flushing
    d floating.

16. Wall ties in straight lengths of cavity walls should be placed vertically every:
    a 300 mm
    b 900 mm
    c 450 mm
    d 350 mm.

17. The purpose of the flashing to a chimney stack is to:
    a prevent birds getting into the roof space
    b act as a lightening conductor
    c connect the roof securely to the stack
    d prevent rainwater passing down the face of the stack into the building.

18. Wall ties are built into cavity walls to:
- **a** prevent the penetration of moisture
- **b** stabilize both walls
- **c** allow the walls to move
- **d** keep the correct cavity width.

19. A flue liner is used in order to:
- **a** maintain a parallel flue
- **b** absorb condensation
- **c** prevent heat loss
- **d** resist the affect of acid attack to the stack.

20. Air bricks are placed in cavity walls in order to:
- **a** increase the sound insulation
- **b** ventilate the air space under the timber floors
- **c** increase thermal insulation
- **d** ventilate the cavity.

21. Which part of the building regulations relates to chimney breasts and stack?
- **a** Part L
- **b** Part B
- **c** Part F
- **d** Part G.

22. Vertical and horizontal DPCs are placed at openings to:
- **a** give additional support to the wall
- **b** prevent moisture entering the inner leaf
- **c** prevent noise entering
- **d** keep leaves apart.

23. The section of a chimney breast that takes the discharge to the external air is called the:
- **a** flue
- **b** chimney
- **c** hearth
- **d** superimposed hearth.

24. Timber cavity battens are used to:
- **a** keep the cavity the correct distance apart
- **b** stop water entering the cavity
- **c** keep the cavity clean
- **d** protect the wall ties.

25. Wall ties must be firmly bedded in each leaf for a minimum of:
- **a** 25 mm
- **b** 50 mm
- **c** 75 mm
- **d** 100 mm.

Now check your answers from the grid

Q 1; d	Q 6; a	Q 11; d	Q 16; c	Q 21; a
Q 2; a	Q 7; c	Q 12; a	Q 17; d	Q 22; b
Q 3; d	Q 8; d	Q 13; a	Q 18; b	Q 23; a
Q 4; c	Q 9; b	Q 14; c	Q 19; d	Q 24; c
Q 5; b	Q 10; d	Q 15; b	Q 20; b	Q 25; b

Section 4.2

1. What is the depth of the fireplace recess?
   a   215 mm
   b   150 mm
   c   500 mm
   d   338 mm.

2. Why is the cavity below the ground level filled with a weak concrete mix?
   a   to prevent the DPC sagging
   b   for thermal insulation
   c   prevent the cavity wall collapsing inwards when backfilling
   d   to give a stable base for the blockwork above ground level.

3. What is the minimum width of the fireplace jambs?
   a   200 mm
   b   500 mm
   c   300 mm
   d   450 mm.

4. The use of a wall plate on top of a wall is to:
   a   help distribute the load
   b   increase the roof/floor space
   c   reduce the number of roof/floor joists
   d   form a level bedding surface.

5. The base of the fire opening is called a:
   a   breast
   b   flue
   c   hearth
   d   stack.

6. Water may penetrate to the inner leaf of a cavity wall by:
  a   travelling across the wall ties
  b   travelling across any mortar which may rest on the wall ties
  c   the air within the cavity
  d   travelling across a lintel.

7. The fire opening should be reduced to:
  a   275 mm
  b   175 mm
  c   125 mm
  d   225 mm.

8. The horizontal DPC is placed more than 150 mm above the:
  a   finished floor level
  b   foundation
  c   ground level
  d   oversite concrete.

9. The angle of a flue should not be less than:
  a   30°
  b   90°
  c   45°
  d   15°.

10. A DPC placed under a sill should extend on each side by at least:
  a   75 mm
  b   100 mm
  c   200 mm
  d   150 mm.

11. To ensure the stability of a chimney stack, it should not exceed:
  a   2½ times its least plan dimension
  b   3½ times its least plan dimension
  c   4½ times its least plan dimension
  d   4 times its least plan dimension.

12. A horizontal DPC is placed to stop moisture:
  a   flooding the cavity
  b   running down into the foundation
  c   passing from the outer leaf into the inner leaf
  d   rising from the ground.

13. The minimum lap for a flexible DPC is:
  a   50 mm
  b   75 mm
  c   100 mm
  d   150 mm.

14. A DPC:
    a   provides a level bed
    b   provides a bed for the floor joist
    c   holds back moisture
    d   binds the lower courses together.

15. Moisture is prevented from entering a building at door and window jambs by:
    a   mastic asphalt
    b   extra wall ties
    c   cavity walling
    d   vertical DPC.

16. Which of the following materials could be used as a rigid DPC?
    a   engineering brick
    b   bitumen felt
    c   mastic asphalt
    d   pitch fibre.

17. To prevent damp rising through solid ground floors, the concrete should be placed on:
    a   a damp-proof membrane
    b   pitch fibre
    c   a layer of earth
    d   a layer of hardcore.

18. Flexible DPC should be laid on a mortar bed in order to:
    a   form a better key with the brick
    b   minimize the risk of puncture
    c   increase the resistance to dampness
    d   maintain the thickness of joints.

19. The minimum height of the DPC above ground level should be:
    a   150 mm
    b   75 mm
    c   125 mm
    d   100 mm.

20. How far apart should wall plate anchor ties be placed?
    a   1.5 m
    b   1.2 m
    c   1.0 m
    d   0.5 m.

21. The internal width of a flue liner should be :
    a   522 mm
    b   275 mm
    c   225 mm
    d   550 mm.

22. Which type of DPC would you use in areas that are likely to suffer subsidence?
    a engineering bricks
    b slate
    c polypropylene
    d you do not need any.

23. The approved document L of the building regulations require external cavity walls to give a 'U' value of:
    a 45 W/m²K
    b 0.045 W/m²K
    c 4.5 W/m²K
    d 0.45 W/m²K.

24. A 'U' value is a measurement of:
    a heat gained into the building
    b water penetration into the building
    c sound loss from the building
    d heat loss from the building.

25. The section of a chimney breast that takes the discharge to the external air is called the:
    a flue
    b chimney
    c hearth
    d superimposed hearth.

Now check your answers from the grid

Q 1; d	Q 6; b	Q 11; c	Q 16; a	Q 21; c
Q 2; c	Q 7; d	Q 12; d	Q 17; a	Q 22; c
Q 3; a	Q 8; c	Q 13; c	Q 18; b	Q 23; d
Q 4; a	Q 9; a	Q 14; c	Q 19; a	Q 24; d
Q 5; c	Q 10; d	Q 15; d	Q 20; b	Q 25; a

Section 4.3

1. What is the purpose of wall ties?

2. Sketch three different types of wall ties.

3. What are the spacing of wall ties:
   vertically
   horizontally
   at openings?

4. Why is it important to keep the cavity clean of mortar droppings?

5. What is the purpose of cavity battens?

6. What materials can be used to span an opening?

7. What is the purpose of a weep hole?

8. What is the purpose of an air brick?

9. What precautions should you take when building in full-fill insulation bats?

10. What precautions should you undertake when building in partial-fill insulation bats?

11. Which part of the building regulations controls the construction of chimney breasts?

12. Define the following:
    chimney
    flue
    constructional hearth
    superimposed hearth.

13. What is the width of the flue?

14. How do you reduce the width of the fireplace opening to the size of the flue?

15. Why do you use flue liners?

16. What angle should the flue not exceed?

17. What is put between the flue liner and the brickwork?

18. How do you protect the top of the stack?

19. What is the purpose of a damp proof course?

20. What materials can be used for a damp proof course?

21. Why should you bed a flexible DPC on a thin mortar screed?

22. What is the minimum lap for a DPC?

23. How should you store timber frames on site?

24. How far should wall ties be bedded into each leaf?

25. Why is a DPC placed over a lintel?

Model answers for Section 4.3

1. What is the purpose of wall ties?

   To give stability to cavity walls the inner and outer leaf must be tied together.

2. Sketch three different types of wall ties.

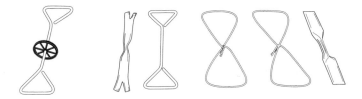

3. What are the spacing of wall ties:

   Vertically – 450 mm
   Horizontally – 900 mm
   At openings – 300 mm but would be reduced to 225 mm to fit in with the blockwork.

4. Why is it important to keep the cavity clean of mortar droppings?

   Mortar droppings can bridge the cavity, allowing moisture to penetrate through to the inside leaf.

5. What is the purpose of cavity battens?

   Cavity battens ensure that the cavity remains free of mortar droppings.

6. What materials can be used to span an opening?

   Concrete or steel lintels, or a combination of both.

7. What is the purpose of a weep hole?

   Weep holes allow any water that has accumulated over an opening to escape to the outer leaf.

8. What is the purpose of an air brick?

   To provide ventilation to food stores, bathrooms or under timber ground floors.

9. What precautions should you take when building in full-fill insulation bats?

   Mortar droppings should not be allowed to accumulate on top of bats.
   All bed and cross joints should be filled fully with mortar.
   Excess mortar should be removed.
   Always follow the manufacturer's instructions.

10. What precautions should you undertake when building in partial-fill insulation bats?

Bats must be securely fixed to the inner leaf.
Bats must be closely jointed.
Bats must be cut accurately to ensure no cold spots occur.
Always store on a flat surface.

11. Which part of the building regulations controls the construction of chimney breasts?

Part L.

12. Define the following:

Chimney – any part of the structure of a building forming any part of a flue other than a flue pipe.
Flue – a passage for conveying the discharge of an appliance to external air.
Constructional hearth – a hearth forming part of the structure of the building.
Superimposed hearth – a hearth not forming part of the structure of a building.

13. What is the width of the flue?

225 mm.

14. How do you reduce the width of the fireplace opening to the size of the flue?

This can be carried out in two ways:
● using a traditional lintel to bridge the opening and the brickwork corbelled in to close the opening.
● using a precast refractory concrete throat unit.

15. Why do you use flue liners?

Flue liners are used to prevent sulphuric acid from getting into the mortar joints and causing the chimney stack to lean.

16. What angle should the flue not exceed?

30°.

17. What is put between the flue liner and the brickwork?

The space between the liner should be filled with a weak mortar mix or insulating concrete.

18. How do you protect the top of the stack?

The top of the stack should be protected with oversailing courses and cement flaunching or concrete capping.

19. What is the purpose of a damp proof course?

To prevent the passage of moisture into the building.

20. What materials can be used for a damp proof course?

Bitumen, pitch polymer, sheet lead and copper, engineering bricks or slate.

21. Why should you bed a flexible DPC on a thin mortar screed?

To prevent the DPC being punctured or damaged.

22. What is the minimum lap for a DPC?

The minimum lap should be 100 mm.

23. How should you store timber frames on site?

Stacked on timber bearers on a level base.
Protected from the elements.
Allow for good air circulation.

24. How far should wall ties be bedded into each leaf?

A minimum of 50 mm into each leaf.

25. Why is a DPC placed over a lintel?

To prevent moisture collecting and eventually penetrating the inner leaf.

ACCESS EQUIPMENT

To complete the assessments in this chapter you will need to know how to:

- identify/take off details of access platforms;
- interpret manufacturers' technical information;
- identify/select type, size and quantities for scaffolds;
- assess and prepare ground conditions;
- position and secure scaffold components;
- stack/store scaffold components.

GLOSSARY OF TERMS

Adjustable base plate – a metal base plate embodying a screw jack, used for adjusting the height of the scaffold at the base.

Base plate – a metal plate with a spigot for distributing the load from a standard or raker or other load-bearing tube (see chart on scaffold fittings, page 97).

Bay – the space between two adjacent standards along the face of a scaffold.

Bay length – the distance between the centre of two adjacent standards measured horizontally.

Board, inside – a board placed between the scaffold and the building on extended transoms, or a hop-up bracket.

Board, retaining – see brick guard.

Board, scaffold – a softwood board combined with others to form access, working platforms and generally used for protective components such as toeboards on a scaffold.

Board clip – a clip for fixing a board to a scaffold tube (see chart on scaffold fittings, page 97).

Box tie – an assembly of tubes and couplers forming a tie for the scaffold by enclosing a feature such as a column.

Brace – a tube placed diagonally with respect to the vertical or horizontal members of a scaffold and fixed to them to afford stability.

Brace, facade or **face** – a brace parallel to the face of the building.

Brace, knee – a brace across the corner of an opening in a scaffold to stiffen the angles or to stiffen the end support of a beam.

Brace, ledger or **cross** – a brace at right angles to the building.

Brace, longitudinal – a brace in the plane of the longer dimension of the scaffold.

Brace, plan – a brace in the horizontal plane.

Bracket, hop-up or **extension** – a bracket to attach (usually to the inside of a scaffold) to enable boards to be placed between the scaffold and the building.

Brick guard – a barrier, usually of coarse mesh, filling the gap between the guard rail and toeboard.

Bridle – a horizontal tube fixed across an opening or parallel to the face of a building to support the inner end of a putlog, transom or tie tube (see drawing on putlog scaffold, page 93).

Butt tube – a short length of tube.

Butting tube – a tube, which butts up against the facade of a building or other surface to prevent the scaffold moving towards that surface.

Coupler, putlog – a coupler used for fixing a putlog or transom to a ledger, or to connect a tube used only as a guard rail to a standard (see chart on scaffold fittings, page 96).

Coupler, right-angle – a coupler used to join tubes at right angles (see chart on scaffold fittings, page 96).

Coupler, sleeve – an external coupler used to join one tube to another (see chart on scaffold fittings, page 96).

Coupler, swivel – a coupler used to join tubes at an angle other than a right angle (see chart on scaffold fittings, page 97).

Decking – the boards or units forming the working platform.

Fittings – a general term embracing components other than couplers.

Gin wheel – a single pulley for fibre ropes attached to a scaffold for raising or lowering materials.

Guard rail – a member incorporated in a structure to prevent the fall of a person from a platform or access way (see drawing of putlog scaffold, page 93).

Hop-up – a wooden access platform constructed with either two or three steps with a surface area of 500 mm × 500 mm and is used by plasterers/painters to reach the ceiling of a room.

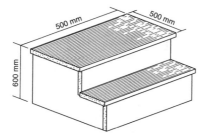

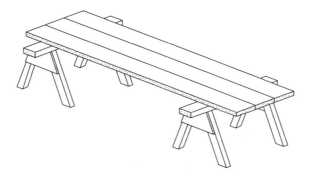

Independent tied scaffold – a scaffold which has two lines of standards, one line supporting the outside of the deck and one on the inside. The transoms are not built into the wall of the building; it is not free standing, but relies on the building for stability.

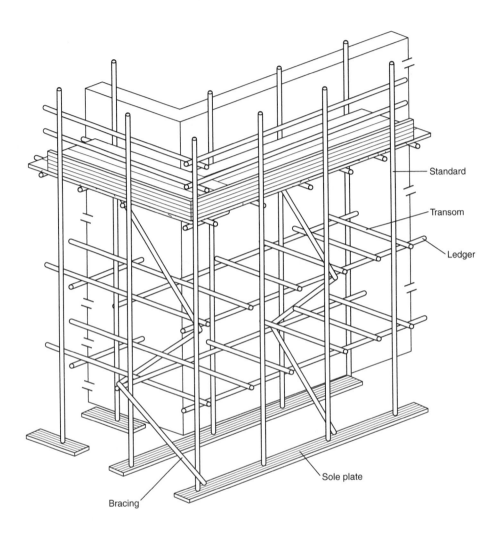

Inside board – a board placed between the scaffold and the building on extended transoms.

Joint pin – an expanding fitting placed in the bore of a tube to connect one tube to another.

Ledger – a longitudinal tube normally fixed to the face of a building in the direction of the larger dimensions of the scaffold. It acts as a support for the putlog and transoms and frequently for tie tubes and ledger braces and is fixed to the adjacent standards.

Lift – the assembly of ledgers and transoms forming each horizontal level of a scaffold.

Lift, foot – a lift erected near to the ground.

Puncheon – a vertical tube supported at its lower end by another scaffold tube and not by the ground.

Putlog – a tube with a blade or flattened end, to rest in the brickwork.

Putlog adaptor – a fitting to provide a putlog blade on the end of a transom (see chart on scaffold fittings, page 96).

Putlog scaffold – a scaffold which has one line of standards to support the outside edge of the deck and utilizes the wall being built or the building to support the inside edge.

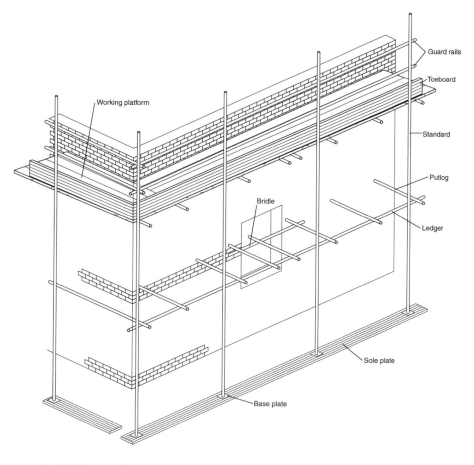

Raker – an inclined load-bearing tube.

Reveal tie – a tube fixed by means of a threaded fitting or by wedging between two opposing surfaces of a structure to form an anchor to which the scaffold may be tied.

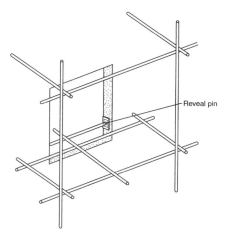

Scaffold – a temporary structure which provides access, or from which persons work, or which is used to support materials, plant or equipment.

Sole board – a timber, concrete or metal spreader used to distribute the load from a standard or base plate to the ground.

Split-head scaffold – a system that provides an adjustable-height staging, suitable for carrying out work at ceiling level.

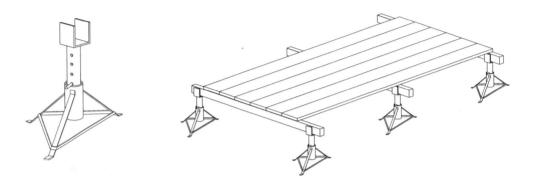

Standard – a vertical or near vertical tube which transfers the load to the ground.

Tie, through – a tie assembly through a window or other opening in a wall.

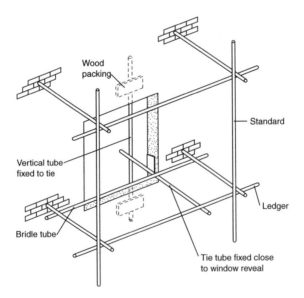

Tie tube – a tube used to connect a scaffold to an anchorage point.

Toeboard – an upstand normally at the outer edge of a platform intended to prevent materials or operative feet from slipping off the platform.

Toeboard clip – a clip used for attaching toeboards to tubes.

Transom – a tube to connect the outer standards to the inner standards or spanning across ledgers to form the support for boards.

Trestle scaffold – intended for light work of short duration, comprises two or more trestles (either folding type or the fixed telescopic type) supporting scaffold or staging boards to form a working platform.

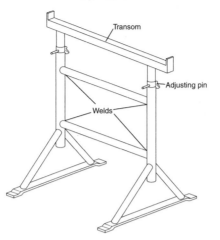

Working platform – the deck from which building operations are carried out.

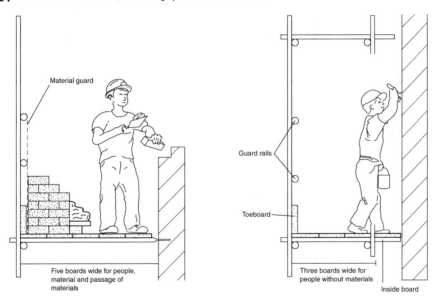

Youngmans staging – constructed of timber and reinforced with steel wire, they are made in various lengths up to 6 m. They are designed to provide access to people using lightweight materials, such as painters.

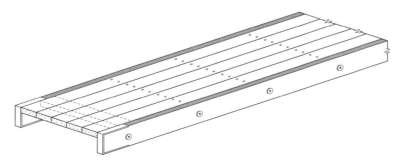

Scaffold fittings chart

Fittings	Description
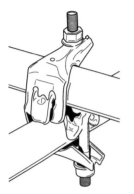	*Putlog coupler* Used to connect a putlog to a ledger.
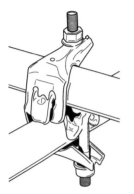	*Double coupler* Used to join two tubes together at right angles.
	*Putlog adaptor* Provides a putlog blade at the end of a transom.
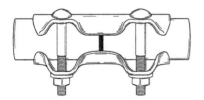	*Sleeve coupler* Used to join two tubes together externally.

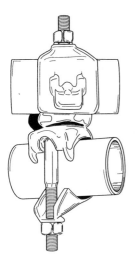

Swivel coupler
Used to join tubes together at an angle other than a right angle.

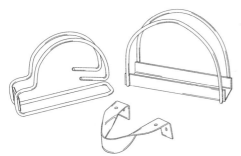

Reveal pin
Inserted into the end of a tube and adjusted to secure scaffolding in window or door openings.

Toeboard clip
Used for attaching toeboards to tubes.

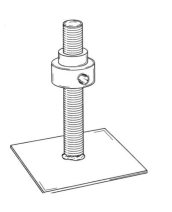

Base plate
A metal plate with a spigot for distributing the load from a standard to the sole board.

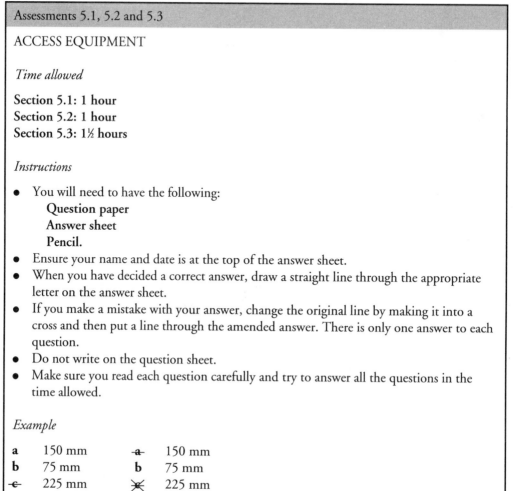

Assessments 5.1, 5.2 and 5.3

ACCESS EQUIPMENT

Time allowed

Section 5.1: 1 hour
Section 5.2: 1 hour
Section 5.3: 1½ hours

Instructions

- You will need to have the following:
    Question paper
    Answer sheet
    Pencil.
- Ensure your name and date is at the top of the answer sheet.
- When you have decided a correct answer, draw a straight line through the appropriate letter on the answer sheet.
- If you make a mistake with your answer, change the original line by making it into a cross and then put a line through the amended answer. There is only one answer to each question.
- Do not write on the question sheet.
- Make sure you read each question carefully and try to answer all the questions in the time allowed.

Example

a	150 mm	~~a~~		150 mm
b	75 mm	**b**		75 mm
~~c~~	225 mm	✷		225 mm
d	300 mm	**d**		300 mm

Section 5.1

1. According to the Lifting Equipment and Lifting Operation Regulations 1998, the minimum width of a scaffold platform for men and materials is:
    a   two boards wide
    b   five boards wide
    c   three boards wide
    d   four boards wide.

2. The projection for a ladder above the top of the platform is:
    a   1.25 m
    b   1.05 m
    c   0.95 m
    d   0.75 m.

3. Cat or crawling ladders are used:
    a within the building
    b in trench excavations
    c between scaffolds
    d for roof work.

4. Which of the following figures is the correct way to store a staging board?

    a

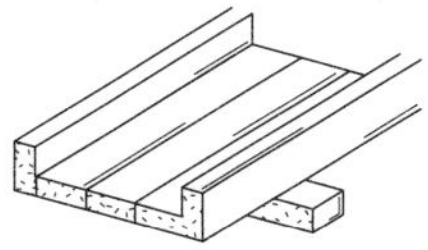

Lay flat upside down
on two supports

    b

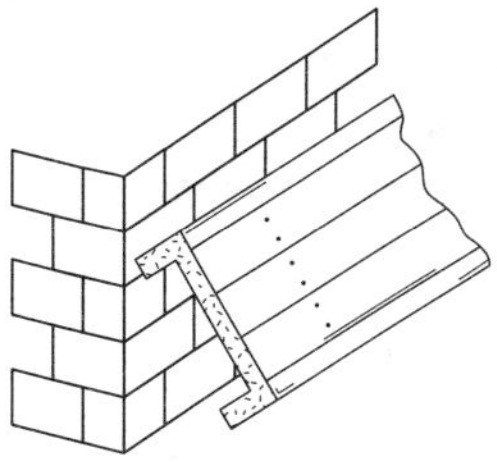

Stand on edge
against a wall

    c

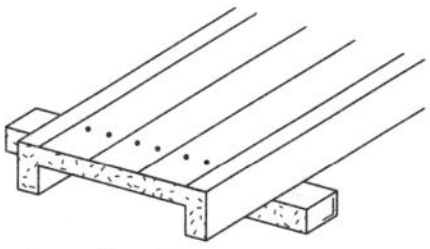

Lay flat the correct way
up on two supports

    d

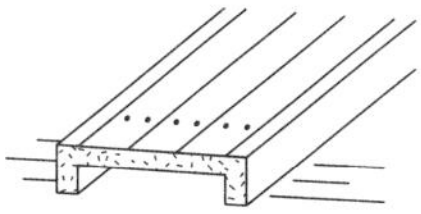

Lay flat on ground
correct way up

5. What is the purpose of a stabilizer on a tower scaffold?
    a to increase the weight of the tower
    b to increase the platform width of the tower
    c to increase the width of the scaffold
    d to increase the height of the scaffold.

6. A tower scaffold, 2.4 m × 1.2 m on plan, is to be erected outside. What is the maximum height without using stabilizers?
    a 4.2 m
    b 3.6 m
    c 35 m
    d 3.9 m.

7. Trestle scaffolds may be taken to a height of:
    a 3.5 m
    b 4.0 m
    c 4.5 m
    d 5.0 m.

8. The correct angle for a ladder is:
    a 75°
    b 65°
    c 70°
    d 85°.

9. The top of a hop-up should be:
   a 500 mm × 400 mm
   b 500 mm × 750 mm
   c 500 mm × 500 mm
   d 450 mm × 450 mm.

10. The maximum height for a free-standing tower scaffold outside should not exceed:
   a 3:1 at least base dimension
   b 3:2 at least base dimension
   c 3½:1 at least base dimension
   d 4:1 at least base dimension.

11. What is the maximum span between two trestles:
   a 3.0 m
   b 2.0 m
   c 2.5 m
   d 3.5 m.

12. A static tower scaffold is one that:
   a moves
   b is used outdoors
   c does not move
   d is used indoors.

13. When using a board to span between two trestles, how far down from the top is it advisable to place the board?
   a a quarter of the way down
   b half way down
   c a third of the way down
   d at the top.

14. The only type of ladder that should be used near overhead electrical cables is:
   a a roof ladder
   b a metal extension ladder
   c a steeplejack ladder
   d a timber ladder.

15. Roof ladders are usually made from:
   a timber
   b aluminium
   c hardwood
   d plastic.

16. The correct treatment for timber ladders is:
  a to paint the stiles but not the rungs
  b should not be painted
  c paint the rungs only
  d paint the stiles and the rungs.

17. The correct angle in which a ladder should rest is:
  a 1 out 4 up
  b 4 out 1 up
  c 1 out 2 up
  d 1 up 3 out.

18. The maximum height of a tower scaffold used inside a building should not exceed:
  a 3:1 least base dimension
  b 3:2 least base dimension
  c 4: least base dimension
  d 3½:1 least base dimension.

19. How high should you raise the stabilizers before moving a mobile tower scaffold?
  a 15 mm
  b 10 mm
  c 20 mm
  d 12 mm.

20. The Construction Regulations state that there should be no gaps between the working platform and the guardrail exceeding:
  a 450 mm
  b 470 mm
  c 500 mm
  d 400 mm.

21. Which part of a ladder protects the user if a rung breaks?
  a reinforcement
  b tie rod
  c safety rod
  d cross tie.

22. What is the name of the two sides of a ladder?
  a rung
  b stile
  c tread
  d tie rod.

23. What is the correct name for the access equipment below:
  a hop-up
  b trestle
  c tower scaffold
  d painters trestle.

24. Guard rails are provided when operatives are liable to fall more than:
  a 1.5 m
  b 2.0 m
  c 1.8 m
  d 2.5 m.

25. A working platform on a tower must be at least:
  a 500 mm wide
  b 300 mm wide
  c 400 mm wide
  d 600 mm wide.

Now check your answers from the grid

Q 1; d	Q 6; b	Q 11; a	Q 16; b	Q 21; b
Q 2; b	Q 7; c	Q 12; c	Q 17; a	Q 22; b
Q 3; d	Q 8; a	Q 13; c	Q 18; d	Q 23; a
Q 4; c	Q 9; c	Q 14; d	Q 19; d	Q 24; b
Q 5; d	Q 10; a	Q 15; b	Q 20; b	Q 25; d

Section 5.2

1. What is the purpose of a toeboard?
  a to strengthen the scaffold
  b to support the guard rail
  c to prevent men falling
  d to prevent materials and tools falling.

2. The scaffold fitting used to connect a cross-brace to a standard is a:
  a swivel coupler
  b universal coupler
  c sleeve coupler
  d double coupler.

3. Sole plates are used in scaffolds:
  a if there are no window openings
  b when the ground bearing strength is suspect
  c to accommodate varying standard lengths
  d where independent scaffolds are required.

4. The purpose of bracing scaffolding is to:
  a increase the weight by using more tubes
  b reduce the weight
  c increase the stability
  d avoid the need for base plates.

5. A double coupler is used on a scaffold:
  a to fasten the transom to ledgers
  b in all positions where strength is needed
  c for all longitudinal bracing
  d for all traverse bracing.

6. What is the minimum number of standards per sole boards?
  a three
  b two
  c one
  d four.

7. What is a putlog scaffold?
  a one which has two rows of standards
  b one which is free standing
  c one which has one row of standards
  d one which has interlocking members.

8. Horizontal members used to support scaffold boards are called:
  a ledgers
  b bracing
  c standards
  d putlog/transom.

9. Where is the blade of the putlog placed into the wall?
  a cross-joint
  b collar joint
  c bed joint
  d open joint.

10. The minimum height of a guard rail is:
    a 955 mm
    b 755 mm
    c 715 mm
    d 915 mm.

11. A ledger is a tube which is:
    a the vertical support of a tube
    b the diagonal support of the scaffold
    c the horizontal member of the scaffold
    d the horizontal member tying in the scaffold.

12. A sleeve coupler:
    a connects two tubes internally
    b connects two tubes externally
    c is a coupler that secures a putlog to the ledger
    d is a coupler that connects a brace to the standard.

13. Toeboards must be provided on all platforms which are higher than:
    a 2.0 m
    b 1.5 m
    c 1.0 m
    d 2.5 m.

14. How deep should the putlog blade be inserted into the wall?
    a 25 mm
    b 100 mm
    c 75 mm
    d 50 mm.

15. What is the minimum size of a sole board?
    a 500 mm × 500 mm
    b 450 mm × 450 mm
    c 250 mm × 250 mm
    d 500 mm × 750 mm.

16. What is the minimum width of a working platform of a trestle scaffold?
    a four boards wide
    b one board wide
    c three boards wide
    d two boards wide.

17. The maximum height of the guardrail is:
    a 1.0 m
    b 1.150 m
    c 1.1 m
    d 1.5 m.

18. The purpose of a bridle is to:
    a   support standards at openings
    b   supports braces at openings ,
    c   support ledgers at openings
    d   support putlogs at openings.

19. Which member of the scaffold is the main load-bearing member?
    a   the putlog
    b   the ledger
    c   the standard
    d   the brace.

20. The toeboard must not be less than:
    a   125 mm high
    b   135 mm high
    c   155 mm high
    d   175 mm high.

21. What is the recommended angle that a raking tube should be positioned?
    a   2 up 1 out
    b   4 up 1 out
    c   3 up 1 out
    d   2 up 3 out.

22. The Lifting Equipment and Lifting Operation Regulations 1998 recommend that scaffold boards should not project beyond the putlog more than:
    a   the width of the board
    b   four times its thickness
    c   300 mm
    d   six times its thickness.

23. What is the name of the tie that is formed around a column?
    a   box tie
    b   physical tie
    c   reveal tie
    d   raker.

24. The maximum height of a tower scaffold, 1.8 m × 1.2 m on plan, used inside a building, should be erected to a maximum height of?
    a   4.2 m
    b   3.6 m
    c   4.5 m
    d   3.5 m.

Now check your answers from the grid

Q 1; d	Q 6; b	Q 11; c	Q 16; d	Q 21; a
Q 2; a	Q 7; c	Q 12; b	Q 17; b	Q 22; b
Q 3; b	Q 8; d	Q 13; a	Q 18; d	Q 23; a
Q 4; c	Q 9; c	Q 14; c	Q 19; c	Q 24; a
Q 5; b	Q 10; d	Q 15; b	Q 20; d	

Section 5.3

1. Name two parts of a ladder.

2. Why should you not paint ladders?

3. What should you do if you find a defect on a ladder?

4. What should you look for on a metal ladder?

5. How should you store ladders?

6. What is the correct angle of a ladder when erected?

7. How should ladders be secured?

8. How much should a ladder extend above a working platform?

9. Name three parts of a stepladder.

10. What should you inspect a stepladder for before using it?

11. How should you use a trestle?

12. What is a split head?

13. What maintenance should be carried out on trestles and split heads?

14. What should you read before erecting a tower scaffold?

15. How high are you allowed to erect a tower scaffold?
    outside
    inside.

16. What precautions should you take when moving a tower scaffold?

17. What is a youngmans staging?

18. Why are scaffolds used?

19. Who should erect a scaffold?

20. What should you look for when inspecting scaffold tubes?

21. How should scaffold tubes be stored?

22. How should scaffold boards be stored?

23. Describe the following fittings:
right angle coupler
swivel coupler
putlog coupler
sleeve coupler
putlog adapter
toeboards clip.

24. How should you take care of scaffold fittings?

25. Describe a putlog scaffold.

26. Describe an independent scaffold.

27. What is the purpose of a scaffold tie?

28. What is the minimum width of a scaffold used for persons only?

29. How wide should the platform be if materials are to be placed onto it?

30. What is the height of a working platform before a guard rail is fitted and at what height should it be above the working platform?

31. Complete the following chart.

Thickness of board (mm)	Maximum span between supports (m)	Minimum overhang (mm)	Maximum overhang (mm)
38			
50			
63			

32. What could happen to a board if it overhangs too much?

Model answers for Section 5.3

1. Name two parts of a ladder.

   Rungs, stiles or tie rods.

2. Why should you not paint ladders?

   Painting can hide defects such as cracks or splitting stiles.

3. What should you do if you find a defect on a ladder?

   Report any defect to your supervisor.

4. What should you look for on a metal ladder?

   Cracks or corrosion.
   Brackets and hooks for condition and anchorage.
   Rubber feet for firmness.

5. How should you store ladders?

   In dry, well ventilated area. Lay flat on rack along the full length.

6. What is the correct angle of a ladder when erected?

   75° or 1 out 4 up.

7. How should ladders be secured?

   A ladder should be lashed or secured at its uppermost resting place and must be tied at the foot. If a ladder cannot be lashed, then you should put something heavy at the base or get someone to stand at the bottom.

8. How much should a ladder extend above a working platform?

   1.05 m.

9. Name three parts of a stepladder.

   Treads, hinges, stiles, restraining rope/tie rod.

10. What should you inspect a stepladder for before using it?

    Treads securely held – no undue wear.
    Cracks or splinters in the stile.
    Hinges securely held.
    Restraining rope – fixed securely, not frayed.

11. How should you use a trestle?

    Should not be painted.
    Never lean outwards or sideways.

Ensure steps are fully extended.
Never work on steps higher than two-thirds the height of the steps.

12. What is a split head?

    This is a system that provides an adjustable height staging suitable for carrying out work at ceiling level.

13. What maintenance should be carried out on trestles and split heads?

    Clean off all mortar droppings etc. and lightly oil all moving parts.

14. What should you read before erecting a tower scaffold?

    The manufacturer's instructions.

15. How high are you allowed to erect a tower scaffold?

    outside – 3 times the least base dimension.
    inside – 3½ times the least base dimension.

16. What precautions should you take when moving a tower scaffold?

    Clear the platform.
    Unlock the brakes.
    Look for overhead hazards and any obstructions at ground level.

17. What are youngmans staging?

    They are designed to provide access for people using lightweight materials in conjunction with trestles. They are made from timber and reinforced with steel wire.

18. Why are scaffolds used?

    To provide a safe access for personnel working at heights.

19. Who should erect a scaffold?

    A trained scaffolder or other competent person.

20. What should you look for when inspecting scaffold tubes?

    Split ends on tubes
    Burred ends
    Excessive corrosion
    Bent tubes
    Angle cut on end.

21. How should scaffold tubes be stored?

    Tubes should be stored on level bearers or in racks in their correct sizes.

22. How should scaffold boards be stored?

Boards should be stacked no more than 20 high, bonded together with short timber battens and placed on level timber bearers, off the ground and under cover if possible.

23. Describe the following fittings:

right angle coupler – joins tubes together at right angles.
swivel coupler – joins tubes at various angles.
putlog coupler – connecting a putlog or transom to a ledger.
sleeve coupler – used to extend one tube to another.
putlog adapter – a fitting that turns a transom into a putlog.
toeboards clip – used to attach toeboards to tubes.

24. How should you take care of scaffold fittings?

Clips should be inspected before use. They should be cleaned and lightly oiled and stored in bins, sorted into types.

25. Describe a putlog scaffold.

This consists of a single row of standards, parallel to the face of the building and set as far away from the building as necessary to accommodate a working platform. A ledger fixed with a right-angled coupler connects the standards. Putlogs are then connected to the ledger with its spade end seated in the bed joint of the adjoining building.

26. Describe an independent scaffold.

This consists of two rows of standards, with each row parallel to the building. The standards are connected to ledgers and fixed with right angle couplers. Transoms are fixed across ledgers to provide a cross-tie and support the working platform. Diagonal bracing is required to give additional support.

27. What is the purpose of a scaffold tie?

Ties should be provided to resist inward and outward movement of scaffolds.

28. What is the minimum width of a scaffold used for persons only?

Three boards wide.

29. How wide should the platform be if materials are to be placed onto it?

Five boards wide.

30. What is the height of a working platform before a guardrail is fitted and at what height should it be above the working platform?

Guardrails should be fitted when the scaffold is at least 2 m high. The guardrail is set at 910 mm above the working platform with a second rail fitted to limit any gap to 470 mm.

31. Complete the following chart.

Thickness of board (mm)	Maximum span between supports (m)	Minimum overhang (mm)	Maximum overhang (mm)
38	1.5	50	150
50	2.6	50	200
63	3.25	50	250

32. What could happen to a board if it overhangs too much?

The board is liable to tip. This is normally called a 'trap'.

SITE DRAINAGE

To complete the assignments in this chapter you will need to know how to:

- identify/take off details of drainage components/materials;
- set out trenches up to 1 m deep;
- lay granular fill to trench bottom;
- lay pipes up to 100 mm diameter;
- construct brick manholes/inspection chambers;
- construct plastic/concrete sectional inspection chambers;
- test drain runs.

GLOSSARY OF TERMS

Benching – this is cement and sand rendering up the sides and above the channel pipes in manholes so as to direct the flow of liquids into the channel itself. The top is usually sloped at an angle of 1:12.

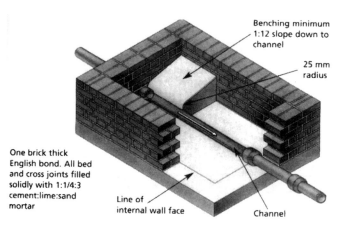

Benching minimum 1:12 slope down to channel

25 mm radius

One brick thick English bond. All bed and cross joints filled solidly with 1:1/4:3 cement:lime:sand mortar

Line of internal wall face

Channel

Bond – this is the arrangement of brick or blocks that form the bond of the wall, usually overlapping to distribute the load.

Boning rods – these are wooden T-pieces used in setting out drains, levelling ground etc. usually used in sets of three.

British Standards – these are rules defining the sizes and properties of materials and their proper use in building.

Cesspool – this is an underground chamber built to receive and store foul water. It is then cleaned out periodically.

Channel pipes – these are half-round pipes found in the base of a manhole or inspection chamber.

Collar joint – a continuous vertical joint running parallel to the face of a wall and is found in walls that are one brick or more in thickness.

Combined system – this is a system of drains by which foul and surface water is conveyed by one set of pipes to the main sewer.

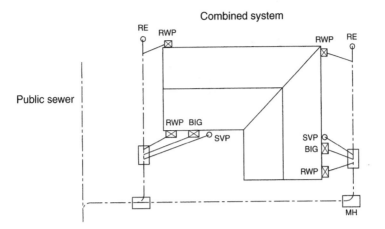

Combined system

Cross-joint – this is the vertical mortar joint between bricks (sometimes called a 'perp').

Drain – a set of pipes that carry away either surface or soil water to a suitable outlet.

Drain rods – a set of rods that connect together with different types of end fittings used in cleaning drains if they get blocked.

Engineering brick – hard dense bricks of regular size used for carrying heavy loads and are impervious to water. Used to build manholes.

Gulley – this is a chamber that is positioned in a drainage system to receive surface or sink water. It may be untrapped or trapped with a minimum 50 mm water seal to prevent gasses being released into the atmosphere.

Back inlet gulley

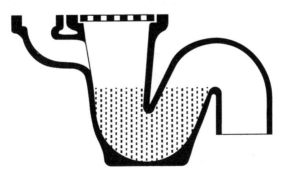

Back (vertical) inlet gully

Inspection chamber – this is a small shallow manhole that allows the drainage system to be inspected or maintained from the ground level (see drawing under 'benching').

Invert – the lowest point on the internal surface of a channel in a drain.

Manhole – a large inspection chamber that allows access for maintenance work to the drainage system from below ground level.

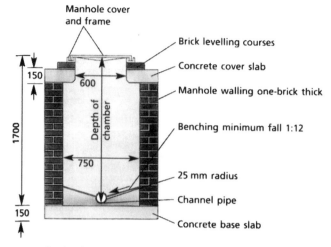

Section through typical manhole looking up stream

Rainwater pipes – a pipe used to carry rainwater from a roof to a drain or water tank.

Rodding eye – this is an inclined shaft constructed at the head of or in a drain run to allow the clearance of any blockages that occur, by using drain rods.

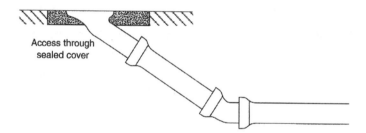

Separate system – this is a system of drains by which both foul and surface water are carried in two separate sets of drains or sewers.

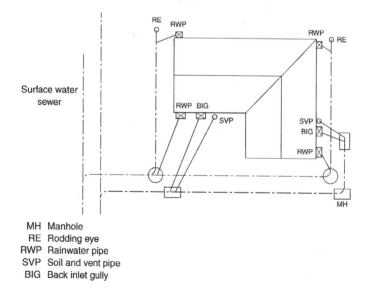

MH Manhole
RE Rodding eye
RWP Rainwater pipe
SVP Soil and vent pipe
BIG Back inlet gully

Sewer – this is a pipe used to carry off soil liquids for treatment at sewage works.

Site datum – this is a fixed reference point from which all levels are taken. It may take the form of a temporary peg set in the ground or it could be something permanent, such as a manhole cover frame, doorstep of adjacent building, roadside kerbs etc.

Soakaway – a means of disposing surface water so that it will percolate into the subsoil.

Soil drain – a pipe used to carry wastewater from above ground WCs etc. to the sewage works.

Soil pipe – a pipe used to carry waste water from WCs etc. to the drain.

Storm water – rainwater discharged from a catchment area, such as a road or paved area, as a result of a storm.

Surface water – the run off of natural water from surfaces including paved areas or roofs etc.

Surface water pipe – drain pipes for carrying away rain or surface water only to a suitable outlet, such as a stream, river or lake.

Trap – this is a water seal found in a gulley to prevent the backflow of obnoxious gasses into the building, must be a minimum of 50 mm.

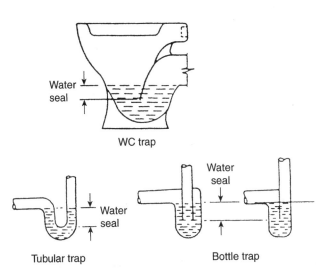

Ventilation pipe – the top part of a soil pipe above the highest branch which carries on above the eaves of the roof, the pipe being left open to allow the escape of foul air from the drains.

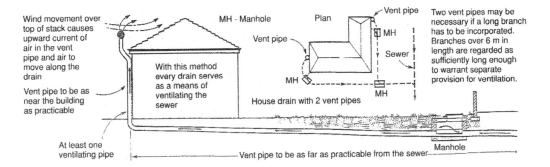

Waste pipe – a pipe carrying the water discharged from a bath, sink, washbasins etc.

Assessments 6.1, 6.2 and 6.3

SITE DRAINAGE

Time allowed

Section 6.1: 1 hour
Section 6.2: 1 hour
Section 6.3: 1½ hours

Instructions

- You will need to have the following:
    Question paper
    Answer sheet
    Pencil.
- Ensure your name and date is at the top of the answer sheet.
- When you have decided a correct answer, draw a straight line through the appropriate letter on the answer sheet.
- If you make a mistake with your answer, change the original line by making it into a cross and then put a line through the amended answer. There is only one answer to each question.
- Do not write on the question sheet.
- Make sure you read each question carefully and try to answer all the questions in the time allowed.

Example

a	150 mm	a̶	150 mm
b	75 mm	**b**	75 mm
c̶	225 mm	✚̶	225 mm
d	300 mm	**d**	300 mm

Section 6.1

1. What is the name of the bend shown below?
    a long bend
    b medium bend
    c short bend
    d taper bend.

2. Which of the following bends is a rest bend?

a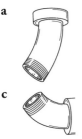

b

c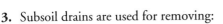

d

3. Subsoil drains are used for removing:
  a rainwater that has seeped into the ground
  b foul water
  c surface water
  d both foul and surface water.

4. What is the purpose of a soakaway?
  a it allows access for cleaning and maintenance work to the drainage system
  b it allows rain water to drain away into the subsoil
  c it collects silt etc. to be removed at a later date
  d it allows for adequate ventilation of the drainage system.

5. Surface water is defined as:
  a water collected from baths, sinks or basins
  b water collected from urinals, WCs etc.
  c water that has soaked into the ground
  d the natural run-off of water from the ground.

6. One method of ensuring the correct fall for a drain is to use a:
  a tapered board
  b tamping board
  c barge board
  d poling board.

7. A wooden T-piece used in drainage is a:
  a site rail
  b boning rod
  c template
  d poling board.

8. Drains are said to be self-cleansing when they are:
  a laid in straight lengths
  b have manholes well spaced
  c are laid to the appropriate fall
  d are well ventilated.

9. Clay drain pipes which pass under buildings should be surrounded by:
    a 150 mm layer of concrete
    b 200 mm layer of broken bricks
    c 100 mm layer of pea gravel
    d 125 mm layer of sand.

10. A tapered straight edge is used in conjunction with a:
    a sight rail
    b foundation brickwork
    c stepped foundation
    d drain pipe.

11. Which of the figures below show a rigid joint?

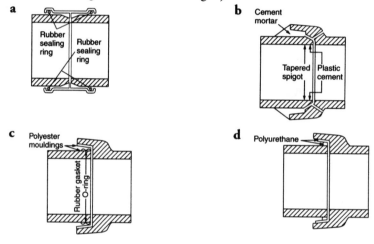

12. The drainage fitting which is used at the bottom of a soil pipe is a:
    a gulley trap
    b rest bend
    c rain water shoe
    d back-inlet gulley.

13. Flexible jointed pipes should be bedded on:
    a clinker ash
    b river sand
    c pea gravel
    d crushed bricks.

14. Backfilling and compaction of trenches should be carried out in layers of:
    a 450 mm
    b 200 mm
    c 300 mm
    d 150 mm.

15. In open timbering, the poling boards are retained in position by using:
    a puncheons
    b lacing
    c struts
    d walings.

16. A timbering system is kept secure by:
    a nailing the members together
    b screwing the members together
    c using lacings
    d using wedges.

17. A back-inlet gulley can be positioned in a drain to provide a:
    a connection between soil pipe and drain
    b rodding arm to clear the drain if it gets blocked
    c connection between rainwater pipes and drain
    d suitable location for a fresh air inlet.

18. A drain 8.5 m long is laid to a fall of 1:40. What is the total fall of the drain to the nearest millimeter?
    a 0.035 m
    b 2.212 m
    c 0.212 m
    d 3.4 m.

19. What is the most suitable type of brick to build a manhole?
    a class B engineering brick
    b clay brick
    c sand/lime brick
    d glass brick.

20. The partially separate system has one major disadvantage in that it:
    a only flushes the system from time to time
    b uses less manholes
    c uses less pipes
    d collects water from paved areas.

21. French drains is a common term for:
    a storm drains
    b foul drains
    c house drains
    d surface water drains.

22. An adjustable trench prop can be used in timbering to trenches in place of a:
    a runner
    b waling
    c poling
    d strut.

23. Timbering in trenches should be inspected daily by a:
   a   drains inspector
   b   clerk of works
   c   building inspector
   d   site foreman.

24. Barriers must be placed around a trench when its depth exceeds:
   a   1.0 m
   b   1.5 m
   c   2.0 m
   d   2.5 m.

25. Poling boards are used to:
   a   support trench sides
   b   prop up the building
   c   provide formwork for concrete
   d   support centres for arches.

Now check your answers from the grid

Q 1; c	Q 6; a	Q 11; b	Q 16; d	Q 21; d
Q 2; d	Q 7; b	Q 12; b	Q 17; c	Q 22; d
Q 3; a	Q 8; c	Q 13; c	Q 18; c	Q 23; d
Q 4; b	Q 9; a	Q 14; d	Q 19; a	Q 24; b
Q 5; d	Q 10; d	Q 15; c	Q 20; a	Q 25; a

Section 6.2

1. The main reason for benching a manhole is to:
   a   protect the base
   b   strengthen the wall
   c   provide self-cleansing
   d   protect the channel.

2. Inspection chambers are provided in drainage systems in order to allow:
   a   the condition of the effluent to be examined at regular intervals
   b   adequate ventilation of the drainage system
   c   access for inspection, cleaning and maintenance work
   d   silt to accumulate for periodic removal.

3. Where is benching used?
   a to form a bed for the drain
   b to cover stoneware pipes
   c between two drain pipes
   d in the base of a manhole.

4. The typical size of a poling board used for timbering to trenches is:
   a 50 mm × 25 mm
   b 220 mm × 37 mm
   c 150 mm × 100 mm
   d 220 mm × 100 mm.

5. When a pipe passes through a wall, what type of joint is placed on either side of the wall?
   a a flexible joint
   b a rigid joint
   c no joints
   d a stiff joint.

6. Gulley inlets to drains should have a minimum water seal of:
   a 32 mm
   b 100 mm
   c 75 mm
   d 50 mm.

7. The ball test on drain pipes is used as a check for:
   a well-worn pipes
   b lipping of the pipe joints
   c any leakage in the pipes
   d correct amount of fall.

8. The water test requires a head of water at the highest point not exceeding:
   a 1 m
   b 1.4 m
   c 1.5 m
   d 1.2 m.

9. Step irons used in manholes should be spaced at a maximum vertical distance of:
   a 350 mm
   b 225 mm
   c 200 mm
   d 300 mm.

10. A saddle fitting is used to:
    a reduce the rate of flow
    b connect a drain to a sewer
    c form outlets from manholes
    d inspect long lengths of drains.

11. A separate system of drainage is one that:
    a   conveys foul and surface water in two sets of pipes
    b   conveys foul and surface water in one set of pipes
    c   conveys foul water only
    d   conveys surface water only.

12. To prevent displacement of struts in trench timbering the following are used:
    a   waling boards
    b   puncheons
    c   lipping pieces
    d   page wedges.

13. A private sewer is a length of pipe serving:
    a   one household and owned by them
    b   two or more householders and owned by them
    c   two or more householders and owned by the local authority
    d   one householder and owned by the local authority.

14. The maximum distance between inspection chambers must not exceed:
    a   90 m
    b   45 m
    c   35 m
    d   50 m.

15. Ventilation should be provided in a drain or private sewer at the:
    a   highest and lowest point
    b   lowest point
    c   every manhole
    d   highest point.

16. A combined system of drainage is one that:
    a   conveys foul and surface water in two separate sets of pipes
    b   conveys foul and surface water in one set of pipes
    c   conveys foul water only
    d   conveys surface water only.

17. A horizontal member used in trench timbering is a:
    a   waling board
    b   poling board
    c   foot prop
    d   puncheon.

18. A water test is carried out on a completed length of drain to ensure:
    a   there is a self-cleansing gradient
    b   the joints are watertight
    c   the drainage system works
    d   there is no blockage in the drains.

19. The figure below shows a trench adjacent to a foundation wall. The trench should be filled with:

   a   concrete to the upper side of the foundation
   b   concrete to the level of the underside of the foundation
   c   hardcore up to the level of the underside of the foundation
   d   concrete up to ground level.

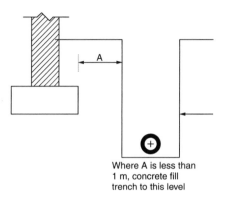

Where A is less than
1 m, concrete fill
trench to this level

20. When setting out invert levels using a dumpy level, the first reading at the lowest point of the drain is 8.34. The drain is 50 m long and the gradient is 1:60, what should be the reading at the highest point of the drain?

   a   0.04
   b   8.432
   c   7.51
   d   9.17.

21. The benching on a manhole should be laid to a slope of:

   a   1:60
   b   1:40
   c   1:12
   d   1:20.

22. Which drainage system is the most expensive?

   a   combined system
   b   separate system
   c   part separate system
   d   cesspool.

23. A foul sewer is normally designed to carry:

   a   storm water
   b   a combination of foul and storm water
   c   an accumulation of sewer gasses
   d   foul water.

24. Planking and strutting is provided in trenches to:
    a enable the trench to be kept open to the elements
    b prevent the collapse of the sides of the trench
    c enable materials and spoil to be placed along the edge of the excavation
    d allow for mechanical excavation.

25. The drainage system shown below is called a:
    a separate system
    b subsoil system
    c combined system
    d part separate system.

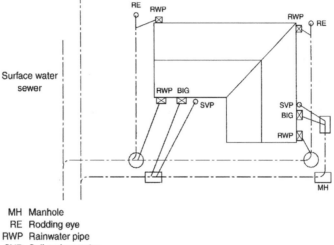

MH Manhole
RE Rodding eye
RWP Rainwater pipe
SVP Soil and vent pipe
BIG Back inlet gully

Now check your answers from the grid

Q 1; c	Q 6; d	Q 11; a	Q 16; b	Q 21; c
Q 2; c	Q 7; b	Q 12; c	Q 17; a	Q 22; b
Q 3; d	Q 8; c	Q 13; b	Q 18; b	Q 23; d
Q 4; b	Q 9; d	Q 14; a	Q 19; b	Q 24; b
Q 5; a	Q 10; b	Q 15; d	Q 20; c	Q 25; a

Section 6.3

1. Define the following terms:
drain
private sewer
public sewer
surface water
invert level.

2. Describe the following types of drainage systems:
combined system
separate system.

3. List the advantages of the combined system over the separate system.

4. How do you check the fall of the trench by using boning rods?

5. What is the purpose of a saddle?

6. Why is it important to control the depth of the excavation?

7. Describe how you would use a string line for laying the drains.

8. What precautions should be taken when backfilling the trench?

9. What precautions should be taken when using a water test in a waterlogged trench?

10. How can you test a drain run for obstructions?

11. If the ball does not reach the other end, what should you do?

12. Describe the following tests:
water test
air test.

13. What is the purpose of a manhole?

14. What is the difference between a manhole and an inspection chamber?

15. What type of brick is used for building manholes and what is the recommended mortar mix used?

16. What is the purpose of the benching?

17. With the aid of a neat sketch, show the position of step irons.

18. Why is it important to lay drains to have a self-cleaning velocity?

19. What are the recommended falls for the following?
100 mm diameter pipes
150 mm diameter pipes.

20. Why do the drains need to be ventilated?

21. How can the drains be ventilated?

22. The invert level at the top of a length of drains 50 m long is 2.5 m below datum and the fall is 1:40. Find the invert level of the bottom inspection chamber.

23. Define the following terms:
poling board
waling board
puncheons
strut.

24. What are the main reasons for supporting the sides of a trench?

Model answers for Section 6.3

1. Define the following terms:

Drain – this includes all pipes, fittings etc. laid to remove soil, waste or surface water from one building.
Private sewer – this is a drainage system that serves two or more properties and they are jointly responsible for its maintenance and repair.
Public sewer – any length of drain that is not covered by the above and is the property of the Local Authority who is responsible for all maintenance work.
Surface water – the run-off of natural water from surfaces including paved areas, roofs and unpaved areas.
Invert level – the vertical distance from the top of the manhole to the bottom of the main drain channel.

2. Describe the following types of drainage systems:

Combined system – this is one large system where both surface and soil water lead into one main sewer. Both soil and surface water is led to the sewerage treatment works.
Separate system – this system has two sets of pipes, one for soil water and one for the surface water. The soil water is taken to the sewage works for treatment, whereas the surface water is led to a river etc.

3. List the advantages of the combined system over the separate system.

In the combined system there is only one set of pipes to maintain.
There is no chance of connecting to the wrong drain/sewer.
The layout is simple and straightforward.
In the separate system the soil sewer is not flushed with rainwater; therefore, great care must be taken to ensure a self-cleansing velocity is maintained.

4. How do you check the fall of the trench by using boning rods?

Use of a set of three boning rods can control the fall of the trench. One rod is fixed or held at each end of the excavation and the third is placed on the trench bottom at intervals along the trench. One person then sights through the other two rods at either end.

5. What is the purpose of a saddle?

This is used to connect a length of drain to the top of an existing sewer or main drain.

6. Why is it important to control the depth of the excavation?

This saves labour in 'bottoming out' the trench by hand or using extra bedding material in 'filling in'.

7. Describe how you would use a string line for laying the drains.

The string line is set up in the bottom of the trench to the correct levels. The pipes are then laid under the line with the tops just touching the string line.

8. What precautions should be taken when backfilling the trench?

All bedding materials should be carefully compacted around the pipe in 75 mm layers until the top of the pipe, and then the excavation should be completed in 300 mm layers.

9. What precautions should be taken when using a water test in a waterlogged trench?

Coloured powder should be placed in the testing water, and any leaks will then be quickly noticed.

10. How can you test a drain run for obstructions?

By using the ball test. A solid rubber ball is inserted into the top end and should roll freely down the invert of the drains.

11. If the ball does not reach the other end, what should you do?

Insert drain rods into the pipeline until they touch the ball; the rods are then removed and laid alongside the drain to show the position of the blockage, which should be corrected.

12. Describe the following tests:

Water test:
- inspect pipe line for possible damage;
- fix apparatus to provide head of water to top of drain;
- fix drain plug or air bag to lower end and to any other branches;
- fill drain with water;
- inspect drains for obvious leaks;
- top up drain after 1 hour due absorption into pipes;
- after 30 minutes measure the amount of water loss; this should not exceed 1 litre per hour.

Air test:
- fix drain plug to lower end and any other branches;
- pump air into pipe line;
- 100 mm of pressure should be indicated on the 'U' gauge;
- the pressure should not fall to less than 75 mm during a period of 5 minutes.

13. What is the purpose of a manhole?

To provide access to a sewer or drain at the following:
- a junction;
- a change in the size of the pipes;
- a change of gradient or direction;
- on long lengths at a maximum of 90 m apart.

14. What is the difference between a manhole and an inspection chamber?

Inspection chamber – This is a shallow chamber, governed in size by the cover and frame. Workmen operate from the ground when rodding a drain from it.
Manhole – This is a deeper chamber and must allow enough headroom for men to work safely inside it. The access shaft should be at least 55 mm wide and step irons should be provided in deep manholes.

15. What type of brick is used for building manholes and what is the recommended mortar mix used?

Most authorities require the use of class B engineering bricks laid in 1:4 sand and sulphate-resisting cement.

16. What is the purpose of the benching?

Benching is a raised concrete surface between channels and walls. It must slope sufficiently to direct overflowing sewage back into the channels but be flat enough to provide a safe foothold when rodding inside the chamber.

17. With the aid of a neat sketch, show the position of step irons.

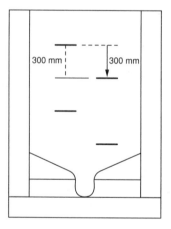

18. Why is it important to lay drains to have a self-cleaning velocity?

To enable the drain runs to be efficient it is important to ensure they have a self-cleansing velocity (known as McQuires rule). If the gradient is too low the speed of the flow will be insufficient, resulting in solids being left behind. If the gradient is too steep the flow of water will be too quick resulting with the same outcome.

19. What are the recommended falls for the following:

100 mm diameter pipes – 1:40 1 metre in 40 m.
150 mm diameter pipes – 1:60 1 metre in 60 m.

20. Why do the drains need to be ventilated?

Drains need to be ventilated to prevent a build-up of sewage gasses in the drainage system. This is done by drawing air into the lowest part of a drain using fresh air inlets (FAIs) in the lowest inspection chamber or by drawing air up through the highest ventilation pipe of the drainage system.

21. How can the drains be ventilated?

By providing a soil and vent pipe at the highest point of the drainage system.

22. The invert level at the top of a length of drains 50 m long is 2.5 m below datum and the fall is 1:40. Find the invert level of the bottom inspection chamber.

$$\text{total fall} = \frac{\text{actual distance}}{\text{given fall}}$$

$$= \frac{50}{40}$$

$$= 1.25 \text{ m}$$

Invert level of bottom inspection chamber = 2.5 − 1.25 = 1.25 m.

23. Define the following terms:

Poling board – short vertical member in direct contact with the trench sides.
Waling board – long horizontal members placed up against the poling boards or walings to prevent their movement.
Puncheons – vertical members fastened between walings to prevent their downward movement.
Struts – strong horizontal members placed up against the poling boards to prevent their movement.

24. What are the main reasons for supporting the sides of a trench?

To provide a safe working environment.
To prevent disruption of work due to collapse of the trench sides.
To prevent damage to adjacent property while the work is carried out.

DECORATIVE BONDING

> # To complete the assignments in this chapter you will need to know how to:
>
> - interpret instructions;
> - plan, organize own work;
> - prepare and install materials and components;
> - construct complex walling details.

GLOSSARY OF TERMS

Axed arch – an arch formed of voussoirs that are cut to a wedge shape with a club hammer and bolster chisel.

Band course – a single course of bricks forming a decorative contrast of bricks.

Batter board – a template used in setting out the batter of a wall.

Bull nose – special shaped bricks with a curved surface joining two adjacent faces (see page 137).

Bullseye – a circular opening in brickwork formed with a completing ring of voussoirs.

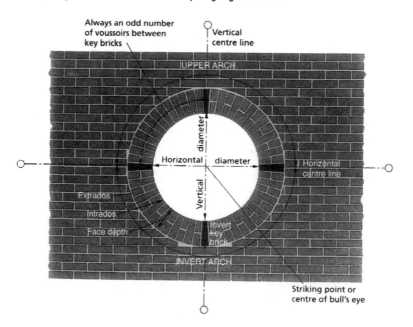

Always an odd number of voussoirs between key bricks

Vertical centre line

UPPER ARCH

Horizontal diameter

Vertical diameter

Horizontal centre line

Extrados
Intrados
Face depth

Invert key brick

INVERT ARCH

Striking point or centre of bull's eye

Camber – a very flat upward curve.

Camber arch – an arch with a slight upward curvature.

Cant – shaped brick with a splayed surface joining to adjacent faces (see page 137).

Corbel – a feature of projecting bricks from the face of the wall.

Dentil course – this is a decorative feature found in walls, it is formed with alternate headers either projecting from the wall or recessed. Sometimes a dentil course is used between oversailing courses.

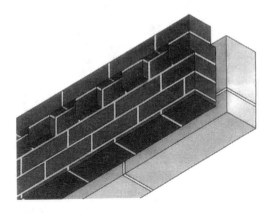

Diaper – a decorative pattern of diagonal intersections or diamond shapes produced by contrasting coloured bricks.

Dogleg – shaped angle brick (see page 137).

Dog toothing – this is another means of forming a decorative effect on the face of brick walls; the dog toothing is obtained by setting out each brick at 45° to the wall face. These may be projecting, flush or recessed from the wall face.

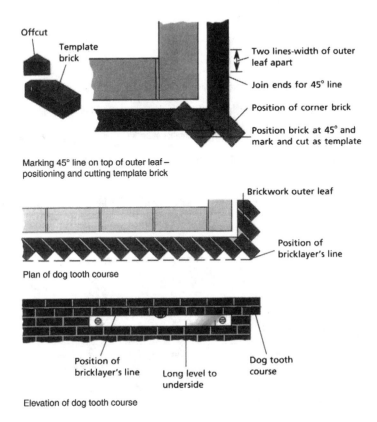

Offcut

Template brick

Two lines-width of outer leaf apart

Join ends for 45° line

Position of corner brick

Position brick at 45° and mark and cut as template

Marking 45° line on top of outer leaf – positioning and cutting template brick

Brickwork outer leaf

Position of bricklayer's line

Plan of dog tooth course

Position of bricklayer's line

Long level to underside

Dog tooth course

Elevation of dog tooth course

Dutch bond – the bonding arrangement consist of alternate course of headers and stretchers; a three-quarter bat is used on the quoin on every stretcher course to provide three-quarter brick lap, and a header is introduced into alternate stretcher courses.

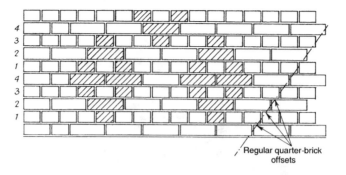

Regular quarter-brick offsets

Efflorescence – a white powdery deposit on the face of brickwork due to the drying out of soluble salts washed from the brick.

English cross-bond – this bonding arrangement consists of alternate courses of headers and stretchers, which have quarter-brick laps. On every alternate stretcher course a header is placed next to the quoin stretcher to form the bond.

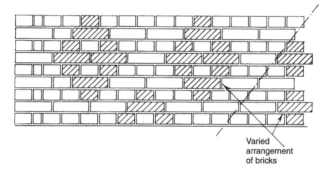

Varied arrangement of bricks

Frost damage – the destructive action of freezing water and thawing ice in saturated materials.

Gauged arch – an arch built of purpose-made or carefully cut wedge-shaped bricks jointed with a non-tapered mortar joint.

Indented quoin – this is done to decorate a quoin by recessing a number of brick courses at the quoin over a short length of walling.

Lime staining – white insoluble deposits on the face of brickwork coming from Portland cement mortars, which have been subjected to severe wetting during setting and hardening.

Monk bond – there are several versions of this bond. The bonding arrangement shown gives a vertical zigzag pattern of contrasting coloured bricks up the entire face of the wall. This bond consists of two stretchers to one header in the same course and a three-quarter is used as the quoin brick.

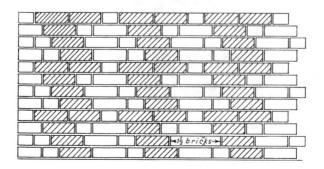

1½ bricks

Plinth (1) – visible projection or recess at the base of the wall.

Plinth (2) – shaped bricks chamfered to provide for reduction in thickness between a plinth and the rest of the wall (see below).

Polychromatic brickwork – decorative patterned work, which features bricks of different colours.

Radial – shaped brick used to build walls curved on plan.

Reference panel – a panel of brickwork built at the start of a contract to set standards of appearance and workmanship.

Retaining wall – a wall built that provides lateral support to higher ground at a change in level.

Rusticated quoin – these are opposite to indented quoins, the brick courses project beyond the face of the quoin.

Sample panel – a panel of brickwork, which may be built to compare materials and workmanship with those of a reference panel.

Specials – bricks of special shape or size used for the construction of particular brickwork features (see below).

Cant bricks

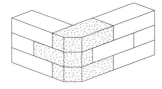
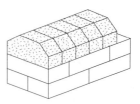

Single cant (left or right hand) –
removes the sharp arris from a wall

Double cant – used for capping walls or
removing the sharp arris from a 1 brick wall

Squint bricks

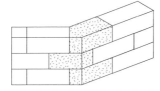
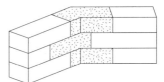

Squint brick (left or right hand) – allows for the
wall to be built at angles of 30°, 45° or 60°

Dog leg brick (left or right hand) – allows for the
wall to be built at angles of 30°, 45° or 60°

Bullnose bricks

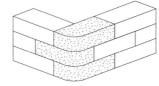
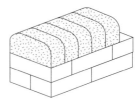

Single bullnose (left or right hand)

Double bullnose

Plinth bricks

Plinth header

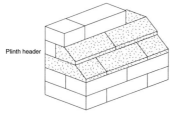
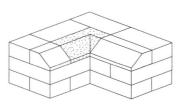

Plinth stretcher

Plinth return

Left hand shown

Squint – a special brick for the construction of non-angled corners (see page137).

Stop – shaped brick to terminate runs of plinths, bullnose or cant bricks.

String course – a long narrow course projecting from the general face of the brickwork.

Sulphate attack – the chemical reaction of soluble sulphates from the ground or certain types of bricks with a chemical constituent of Portland cement which results in expansion of and physical damage to mortar.

Template – full-size pattern, usually of rigid sheet material, used as a guide for cutting or setting out work.

Trammel – timber battens, pivoted at one end, used to set out curved work.

Tumbling in – a sloping feature formed by bricks laid in courses at right angles to the face of the wall.

Assessments 7.1, 7.2 and 7.3

DECORATIVE BRICKWORK

Time allowed

Section 7.1: 1 hour
Section 7.2: 1 hour
Section 7.3: 1½ hours

Instructions

- You will need to have the following:
  Question paper
  Answer sheet
  Pencil.
- Ensure your name and date is at the top of the answer sheet.
- When you have decided a correct answer, draw a straight line through the appropriate letter on the answer sheet.
- If you make a mistake with your answer, change the original line by making it into a cross and then put a line through the amended answer. There is only one answer to each question.
- Do not write on the question sheet.
- Make sure you read each question carefully and try to answer all the questions in the time allowed.

Example

a	150 mm	~~a~~	150 mm
b	75 mm	b	75 mm
~~c~~	225 mm	✗	225 mm
d	300 mm	d	300 mm

Section 7.1

1. When constructing tumbling-in, the ratio of tumbled courses to horizontal courses should be:
   a 3:2 or 3:4
   b 4:2 or 3:2
   c 4:4 or 4:2
   d 3:4 or 6:4.

2. The recess in an indented quoin is usually kept to a maximum of:
   a 32 mm
   b 38 mm
   c 28 mm
   d 56 mm.

3. The elevation shown in the figure below is an example of:
   a English bond
   b Reverse bond
   c English cross-bond
   d Dutch bond.

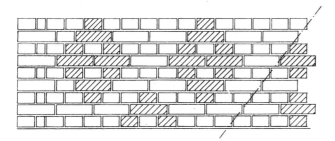

4. A dentil course should not project more than:
   a 75 mm
   b 28 mm
   c 56 mm
   d 12 mm.

5. Square diagonal basket weave panels are set out:
   a below the centre line
   b from the sides of the panel
   c at the base of the panel
   d from the centre of the panel.

6. The equipment required to carry out tumbling-in work is:
   a bevel, square, lines and template
   b square, template, gun and bevel
   c lines, square, template and gun
   d gun, lines, bevel and square.

7. Corbel courses are used to:
   a form string courses
   b increase wall thickness
   c increase stability
   d terminate piers.

8. The amount of tolerance for a trammel pivot rod should be:
   a 2–3 mm
   b 5–6 mm
   c 8–10 mm
   d 10–12 mm.

9. The brick shown below is called a:
   a single bullnose
   b double bullnose
   c cant
   d plinth.

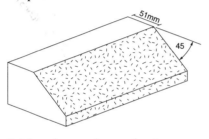

10. Brickwork curved on a plan is set out with a:
   a tracer
   b transom
   c tranversor
   d trammel.

11. Which of the following bricks is used for the construction of an obtuse angle?
   a plinth
   b cant
   c squint
   d king closer.

12. When building curved walls with stretcher bond, the minimum radius is:
   a 5 m
   b 4 m
   c 3 m
   d 2 m.

13. What is the name of the panel that consists of three stretchers stack bond and three soldiers laid adjacent?
   a herringbone
   b stack bond
   c stretcher bond
   d basketweave.

14. When setting out the vertical herringbone panel, where do you set out the first bricks?
   a top of panel
   b centre of panel
   c bottom of panel
   d bottom centre of the panel.

15. Dog toothing is set out at:
   a 30°
   b 45°
   c 25°
   d 60°.

16. An obtuse angle is one that is:
   a more than 90°
   b more than 180°
   c between 90° and 180°
   d less than 90°.

17. What is the name of the decorative bond shown below?
   a diaper bond
   b stretcher bond
   c header bond
   d English cross-bond.

18. The shaped brick shown below is a:
   a squint
   b cant
   c bullnose
   d plinth.

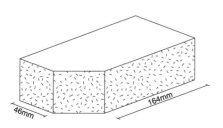

19. A dentil course is a string course in which:
   a the bricks are laid vertical to the horizontal plane and must be kept upright
   b alternate bricks are laid projecting from the face of the wall
   c the bricks are laid at 45° to the face of the wall
   d the bricks form a diagonal pattern in the wall.

20. Name the following type of bond shown below?
   a Dutch bond
   b English cross-bond
   c Monk bond
   d Flemish bond.

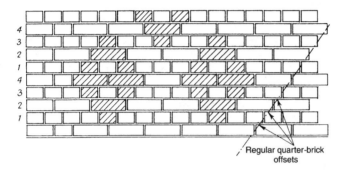

Regular quarter-brick offsets

21. What is the name of the brick panel shown below?
   a herringbone
   b stack
   c basketweave
   d diagonal herringbone.

22. What is the recommended projection on rusticated quoin?
   a 28 mm
   b 38 mm
   c 25 mm
   d 35 mm.

23. When building herringbone panels, the angle can be maintained by using:
    a boat level/45° set square
    b boat level/60° set square
    c gauge rod
    d boat level/gauge rod.

24. Special bricks that are used to build a curved wall are called:
    a headers
    b radials
    c plinths
    d voussoirs.

25. The feature shown below is called:
    a oversail course
    b dentil course
    c string course
    d dog toothing.

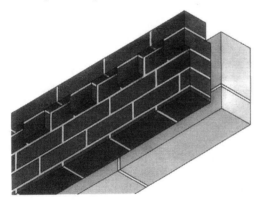

Now check your answers from the grid

Q 1; b	Q 6; d	Q 11; c	Q 16; b	Q 21; b
Q 2; c	Q 7; b	Q 12; c	Q 17; a	Q 22; a
Q 3; c	Q 8; a	Q 13; d	Q 18; b	Q 23; a
Q 4; b	Q 9; d	Q 14; b	Q 19; b	Q 24; b
Q 5; d	Q 10; d	Q 15; b	Q 20; a	Q 25; b

Section 7.2

1. The simplest herringbone panel to set out and build is:
   a   single
   b   double
   c   feather
   d   diagonal.

2. When single herringbone bond is used for rectangular panels, the bricks on each side will be:
   a   similar in length
   b   different in length
   c   opposite to each other
   d   alternate length.

3. The figure shows the first three brick cuts needed for which type of decorative panel?
   a   diagonal basket weave
   b   single herringbone
   c   diagonal herringbone
   d   basket weave.

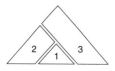

4. Diagonal herringbone is the easiest herringbone panel to set out and build because:
   a   all the work is at 45°
   b   no diagonal cuts are needed
   c   all the cut bricks are the same size
   d   all bricks to be cut can be marked from the same bevel.

5. A string course is built:
   a   under window sills
   b   around the face of a building
   c   over window openings
   d   under arches.

6. Plinth courses are normally used to:
   a   form a decorative effect
   b   form a door/window sill
   c   reduce the thickness of the wall
   d   provide an alternative bond.

7. What is the name of the bond where a header is placed next to the quoins ¾ on every other course of stretchers?
   a   English garden wall bond
   b   English cross-bond
   c   Dutch bond
   d   Monk bond.

8. A purpose made obtuse squint brick has a header face of:
  a 56 mm
  b 75 mm
  c 102 mm
  d 168 mm.

9. The special shaped brick shown below is a:
  a double bullnose
  b single bullnose
  c stepped bullnose
  d cow nose.

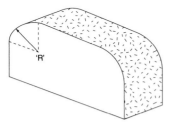

10. In circular work a trammel is used to:
  a set out the radials
  b set out the bond
  c check the level of the brickwork
  d check the alignment of the brickwork.

11. Diaper bond is:
  a diagonal patterns formed in plain walls
  b panels of 3 bricks horizontal and 3 bricks vertical laid alternatively
  c same as above but laid at 45°
  d straight patterns formed in plain walls.

12. The shaped brick shown below is a:
  a squint
  b cant
  c single bullnose
  d plinth.

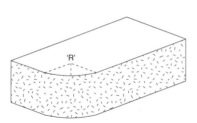

13. With regard to setting out a curved wall, point 'A' in the figure below is called a:
   a stringer
   b striking point
   c trammel point
   d steel point.

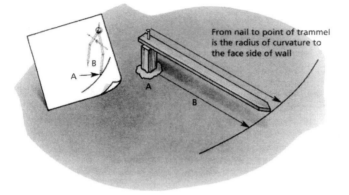

From nail to point of trammel is the radius of curvature to the face side of wall

14. The bond in the panel below is:
   a stretcher bond
   b herringbone
   c stack bond
   d basketweave.

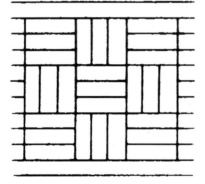

15. The panel shown below is called:
   a diagonal herringbone
   b horizontal herringbone
   c vertical herringbone
   d double vertical herringbone.

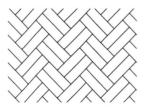

16. A horizontal course of decorative brickwork is called a:
   a string course
   b dog tooth course
   c dentil course
   d rusticated quoin.

17. An acute angle is one that is:
   a more than 90°
   b more than 180°
   c between 90° and 180°
   d less than 90°.

18. Block-bonded corners are those which are built with:
   a 450 × 225 × 100 mm concrete blocks
   b contrasting bonds at the corner
   c contrasting bricks in block patterns at the corner
   d contrasting blocks at the corner.

19. A trammel is a piece of timber which is used to:
   a mark out a curved wall on plan
   b line up a gable wall
   c check corbels in a wall
   d check dog toothing in a wall.

20. The feature shown below is called:
   a oversail course
   b dentil course
   c string course
   d dog toothing.

21. What is the name of the brick panel shown below?
  a horizontal herringbone
  b stack
  c basketweave
  d diagonal herringbone.

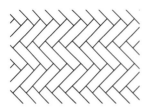

22. What is the name of the bonding arrangement shown below?
  a rusticated quoin
  b indented quoin
  c block bonded quoin
  d block indented quoin.

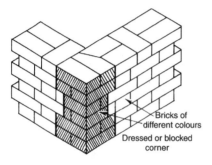

Bricks of
different colours
Dressed or blocked
corner

23. What bond is shown in the figure below?
  a diaper work
  b Monk bond
  c English cross-bond
  d Dutch bond.

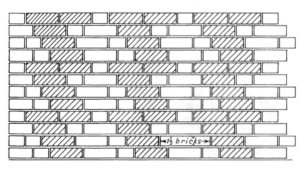

1½ bricks

24. When setting out a curved wall, you need to establish which point first?
  a pivot point
  b striking point
  c radius point
  d trammel point.

25. The name of the template used for building tumbling-in is:
    a   gauge rod
    b   pinch rod
    c   gun
    d   pistol.

Now check your answers from the grid

Q 1; d	Q 6; c	Q 11; a	Q 16; a	Q 21; a
Q 2; a	Q 7; c	Q 12; c	Q 17; d	Q 22; a
Q 3; b	Q 8; a	Q 13; b	Q 18; c	Q 23; b
Q 4; a	Q 9; a	Q 14; d	Q 19; a	Q 24; b
Q 5; b	Q 10; d	Q 15; d	Q 20; d	Q 25; c

Section 7.3

  1. When talking about curved brickwork, what are radials?

  2. What information is needed to enable you to set out a curved wall on plan?

  3. Why is it important to set out the first course accurately when building a curved wall on plan?

  4. When building a bull's-eye, where are the key bricks set out?

  5. Describe how you would set out a curved wall using a trammel.

  6. Describe the difference between a rough arch and an axed arch.

  7. With the aid of a neat sketch, describe how to make a template.

  8. What is the purpose of traversing the arch before it is built?

  9. What is a camber arch?

 10. With the aid of a neat sketch, describe stack bond.

 11. Where do you start setting out a diagonal basket weave panel and why?

 12. List the points to consider when building a basket weave panel.

 13. How do you control the angle when building a diagonal herringbone panel?

14. With the aid of a sketch describe how you would set out a 135° angle.

15. Where would you find an obtuse angle on a building?

16. Where would you use a squint brick and what is its size?

17. What is the purpose of a plinth brick?

18. Using a neat sketch draw the following:
    single cant
    double cant
    dog leg
    squint brick.

19. Describe the purpose of tumbling-in.

20. What is the purpose of a gun or template and what precautions should you take when using one?

21. Complete the following elevation.

    Dutch bond

    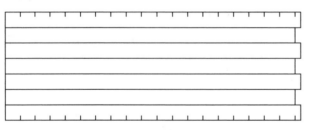

22. List the points to consider when building a dentil course.

23. Describe how you would set out and maintain the correct bond when building dog toothing.

24. Complete the following elevation.

    English cross-bond

    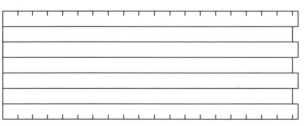

25. What is a rusticated quoin?

Model answers for Section 7.3

1. When talking about curved brickwork, what are radials?

   Radial bricks are special shaped bricks used in constructing curved walls on plan. They allow the cross-joints to be parallel.

2. What information is needed to enable you to set out a curved wall on plan?

   To set out curved walls, find from the drawings the location of the striking point and the radius of the curve.

3. Why is it important to set out the first course accurately when building a curved wall on plan?

   You must set out the first course accurately to the radius mark and check with either a trammel or template to prevent any kinks being built into the wall.

4. When building a bull's-eye, where are the key bricks set out?

   The key bricks should be set out on the horizontal and vertical centre lines, and each quadrant must have an odd number of voussoirs between the key bricks.

5. Describe how you would set out a curved wall using a trammel.

   - A steel rod is placed in concrete at the striking point. Make sure that it is truly plumb and rigidly fixed.
   - Fix the trammel onto the steel rod.
   - On completion of each course, check the alignment of each brick with the trammel. Support the trammel on an elastic band wound round the rod and rolled up for each course.

6. Describe the difference between a rough arch and an axed arch.

   Rough arches have wedged-shaped joints, whereas axed arches have wedged-shaped bricks and parallel joints.

7. With the aid of a neat sketch, describe how to make a template.

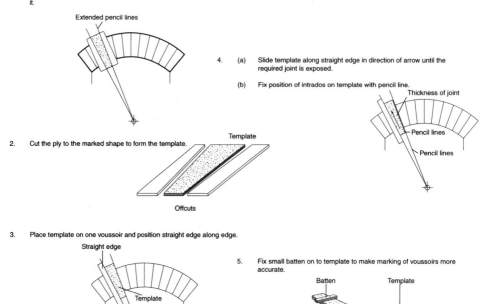

1. Place a piece of ply over the voussoir and mark the voussoir lines onto it.

Extended pencil lines

2. Cut the ply to the marked shape to form the template.

Template

Offcuts

3. Place template on one voussoir and position straight edge along edge.

Straight edge

Template

Centre point

4. (a) Slide template along straight edge in direction of arrow until the required joint is exposed.

   (b) Fix position of intrados on template with pencil line.

   Thickness of joint

   Pencil lines

   Pencil lines

5. Fix small batten on to template to make marking of voussoirs more accurate.

   Batten   Template

8. What is the purpose of traversing the arch before it is built?

To ensure a high-quality arch with tight joints, it is necessary to check that the arch works out with the correct brick sizes and joint sizes.

9. What is a camber arch?

This is also known as a flat arch, it is basically a horizontal band cut from the face of a semi-circular arch. The soffit of the arch is not actually straight, but is given a slight camber of 1 mm in 100 mm.

10. With the aid of a neat sketch, describe stack bond.

This is the simplest of panels and consists of stacking one brick on top of another with no bond.

11. Where do you start setting out a diagonal basket weave panel and why?

At the centre of the panel so that the completed panel is symmetrical (equal size cuts around the edge).

12. List the points to consider when building a basket weave panel.

- Ensure that the bricks are of equal length.
- If a contrasting colour is to be used, ensure a constant colour is maintained.

13. How do you control the angle when building a diagonal herringbone panel?

The angle is controlled by using a 45° set square and a boat level or by using a boat level with an adjustable bubble.

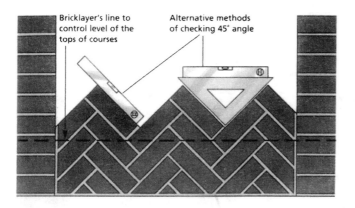

14. With the aid of a sketch describe how you would set out a 135° angle.

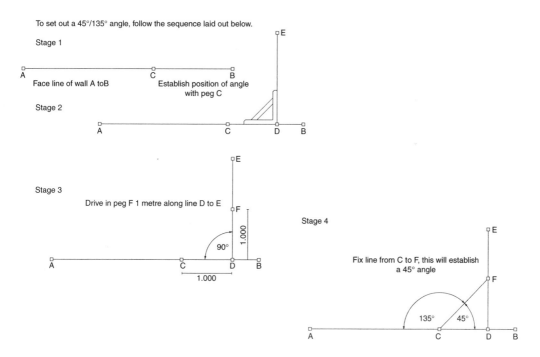

15. Where would you find an obtuse angle on a building?

They are usually found in bay windows or boundary walls.

16. Where would you use a squint brick and what is its size?

At the corner of a wall, which has an obtuse angle. They have a stretcher face of 168 mm and a header face of 56 mm. The angle of the squint is given a number, that is stamped into the face of the frog of the brick.

17. What is the purpose of a plinth brick?

Plinth bricks are mainly used to reduce the thickness of a wall, although they can be used occasionally to increase the thickness.

18. Using a neat sketch draw the following:

Single cant

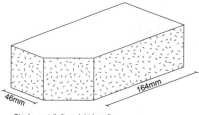

Single cant (left or right hand) –
removes the sharp arris from a wall

Double cant

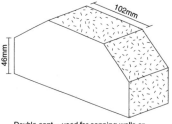

Double cant – used for capping walls or
removing the sharp arris from a 1 brick wall

Dog leg

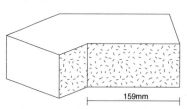

Dog leg brick (left or right hand) – allows for the
wall to be built at angles of 30°, 45° or 60°

Squint brick

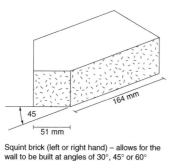

164 mm

45

51 mm

Squint brick (left or right hand) – allows for the
wall to be built at angles of 30°, 45° or 60°

19. Describe the purpose of tumbling-in.

The tumbling-in provides strength and weathering to the top of an attached pier.

20. What is the purpose of a gun or template and what precautions should you take when using one?

A timber gun or template is used to maintain the required slope of the tumbling-in. When using it you should ensure that you press the stem firmly against the plumb face of the attached pier, checking:

- the slope of the face;
- that the tumbling-in courses are at 90° to the template.

21. Complete the following elevation.

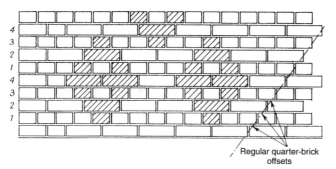

Regular quarter-brick
offsets

Dutch bond

22. List the points to consider when building a dentil course.

- Use solid bricks whenever possible.
- Exposed projecting bricks should be specified to have a frost resistance.
- The eyeline should be formed at the bottom arris of the first course and the top arris of the last course.

23. Describe how you would set out and maintain the correct bond when building dog toothing.

Set out the corner details first and then set out the bond over the whole length. Open or tighten the vertical joints as necessary. Line the cut surface of each dog-tooth brick with the inside face of the external leaf to maintain the 45° angle, but check it with a triangular template.

24. Complete the following elevation.

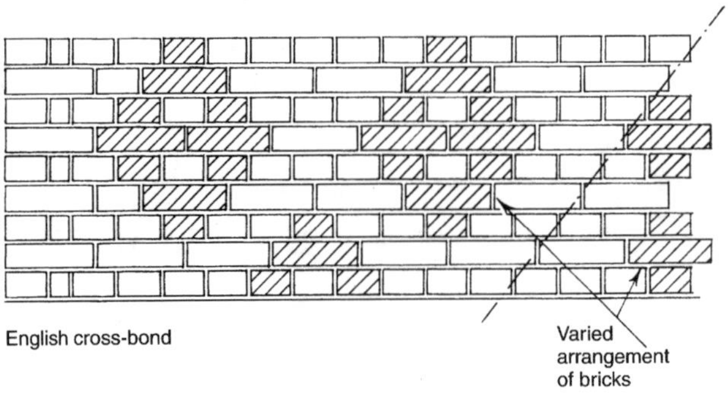

English cross-bond

Varied
arrangement
of bricks

25. What is a rusticated quoin?

This is a decorative feature quoin, built by projecting contrasting bricks from the face of the quoin. They can be in blocks of three or four courses; the dimensions of the rustication can be the same on both faces or the projections can be of alternating lengths.

8

BASIC CALCULATIONS

> # To tackle the assessments in this section you will need to know how to calculate:
>
> - areas of common shapes;
> - volumes of common shapes;
> - quantities of materials to construct walls.

GLOSSARY OF TERMS

Area – a two-dimensional measurement that is multiplied together, such as length × width, and is given in metres squared (m²).

Centre line method – a method of calculating the true length of a wall/foundation etc. by making adjustments for measuring the corners twice.

Circle – a round shape that is measured in square metres.

Circumference – a linear measurement of the outside line of a circle.

Depth – a linear measurement that refers to the vertical measurement of a three-dimensional shape.

Diameter – a straight line from one point of the circumference, passing through the centre of a circle to the opposite side of the circumference.

Length – a linear measurement, such as a straight line. This is usually the longest dimension of a shape and should be written first in a formula.

Linear – a length given in metres.

Pi (π) – a constant value number which is used in determining the area or circumference of a circle. The value of pi is 3.142.

Radius – a straight line running from the centre of a circle to the circumference.

Square – a four-sided two-dimensional figure whose length and width are the same.

Triangle – a three-sided figure which can be used in many calculations, for example when setting out a right angle corner or working out the area of a gable wall.

Volume – a three-dimensional value, e.g. length × width × depth. Volume is measured in metres cubed (m³).

Width – a linear measurement, which often refers to the smaller dimension of a two-dimensional shape.

Useful information

Number of bricks per square metre of walling

Type of wall	Stretcher bond		English bond		Flemish bond		English garden wall bond		Flemish garden wall bond	
	F	C	F	C	F	C	F	C	F	C
½ brick wall	60									
1 brick wall	60	60	90	30	80	40	73	47	67	53
1½ brick wall			90	90	80	100	73	107	67	113
blockwork	10									

F = facings. C = commons.

Area of a circle = πr^2
Area of a square = length × length
Area of a rectangle = length × width
Area of a triangle = ½ base × height
Circumference of a circle = πd
Volume of a cube = length × width × depth

Calculations

A bricklayer needs to know how to work out the area of brickwork to build walls and the cubic metres of concrete for foundations. There are two ways of working out the number of bricks:

1. You can count the number of bricks needed for one course and multiply it by the number of courses in the wall.
   This works well if there are no windows or doors in the wall of for stretcher bond walls, but becomes harder with more complicated shapes.
2. You need to find the surface area of brickwork and then multiply it by the number of bricks per metre square.

EXAMPLES

Below are some worked examples for you to look at. If you set out your calculations in the same style it should help you follow the process through stage by stage.

Example 1

Calculate the area of a square that measures 3 m each side.

$$\text{Area of square} = \text{length} \times \text{length}$$
$$= 3 \times 3 = 9 \text{ m}^2$$

Example 2

Calculate the area of a square measuring 150 mm.

$$\text{Area of square} = \text{length} \times \text{length}$$
$$= 150 \times 150 = 0.0225 \text{ m}^2$$

Example 3

Calculate the area of a rectangle measuring 5 m long by 3 m wide.

$$\text{Area of rectangle} = \text{length} \times \text{width}$$
$$= 5 \times 3 = 15 \text{ m}^2$$

Example 4

Calculate the area of a circle if the radius is 1.5 m.

$$\text{Area of circle} = \pi r^2$$
$$= 3.142 \times 1.5 \times 1.5 = 7.068 \text{ m}^2$$

Example 5

Calculate the cubic area of a foundation measuring 10 m long by 0.5 m wide and 0.2 m deep.

$$\text{Area} = \text{length} \times \text{width} \times \text{depth}$$
$$= 10 \times 0.5 \times 0.2 = 1 \text{ m}^3$$

Example 6

Calculate the circumference of a circle with a diameter of 3 m.

$$\text{Circumference} = \pi d$$
$$= 3.142 \times 3 = 9.42 \text{ m}$$

Example 7

Calculate the area of a triangle that has a base of 3 m and a vertical height of 1.5 m.

$$\text{Area} = \frac{3 \times 1.5}{2}$$
$$= 2.25 \text{ m}^2$$

CENTRE LINE METHOD OF MEASUREMENT

When you are measuring the perimeter of a building with more than one corner, you end up measuring the corners twice. As in the figure shown below, the shaded area of the corner is measured twice.

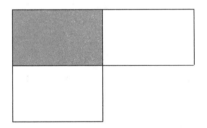

The centre line method overcomes this problem when measuring up walls etc.

Example 8

The ½ brick wall shown below measures 3 m × 2 m externally. Calculate the centre line of the wall.

If you worked out the perimeter by multiplying the length × width it will be:

$$(3 \times 2) + (2 \times 2) = 10 \text{ m}$$

But if you work out the centre line you should deduct the four corners (the width of the trench is 200 mm) thus

$$(3 \times 2) + (2 \times 2) - (4 \times 0.200) = (6 + 4) - 0.800$$
$$= 9.2 \text{ m}$$

Again, if the measurements were given as 1.8 m × 2.6 m internally you would work it out like this:

$$(2 \times 1.9) + (2 \times 2.7) + (4 \times 0.200) = 3.8 + 5.4 + 0.800$$
$$= 9.2 \text{ m}$$

Now you are ready to answer the questions in the following assessments.

Assessments 8.1, 8.2 and 8.3

BASIC CALCULATIONS

Time allowed

Section 8.1: 1 hour
Section 8.2: 1 hour
Section 8.3: 1½ hours

Instructions

- You will need to have the following:
  Question paper
  Answer sheet
  Pencil.
- Ensure your name and date is at the top of the answer sheet.
- When you have decided a correct answer, draw a straight line through the appropriate letter on the answer sheet.
- If you make a mistake with your answer, change the original line by making it into a cross and then put a line through the amended answer. There is only one answer to each question.
- Do not write on the question sheet.
- Make sure you read each question carefully and try to answer all the questions in the time allowed.

Example

a	150 mm	~~a~~	150 mm
b	75 mm	b	75 mm
~~c~~	225 mm	✗	225 mm
d	300 mm	d	300 mm

Section 8.1

1. What is the circumference of a circle if the diameter is 1.5 m?
   a 4.71
   b 3.75
   c 4.50
   d 3.70.

2. What is the area of the triangle shown in the figure below?
   a 0.75 m²
   b 2.0 m²
   c 6.0 m²
   d 0.6 m².

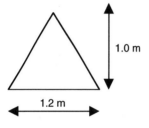

3. What is the area of the triangle shown in the figure below?
   a 3 m²
   b 4 m²
   c 5 m²
   d 6 m².

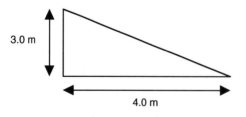

4. If a circle has a radius of 1.75 m, what is its surface area?
   a 9621 m²
   b 96.21 m²
   c 962.1 m²
   d 9.62 m².

5. If a circle has a diameter of 750 mm, what is its surface area?
   a 4.4 m²
   b 0.44 m²
   c 44 m²
   d 444 m².

6. What is the surface area of the circular paved area shown in this figure?
   a   12.56 m^2
   b   1.256 m^2
   c   125 m^2
   d   12 m^2.

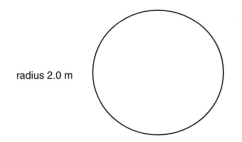

radius 2.0 m

7. Calculate the area of the bull's-eye shown in the following figure.
   a   1963 m^2
   b   19.63 m^2
   c   200 m^2
   d   1.96 m^2.

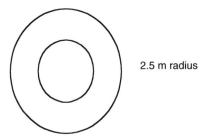

2.5 m radius

8. What is the surface area of the paved area shown in this figure?
   a   592.4 m^2
   b   540 m^2
   c   50.27 m^2
   d   500 m^2.

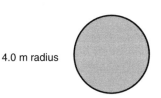

4.0 m radius

9. What is the circumference of the bull's-eye shown in the figure to question 7?
   a   157.0 m
   b   15.71 m
   c   1.57 m
   d   1500 m.

10. What is the circumference of the circle shown in the figure to question 8?

    a   25.13 m

    b   250 m

    c   2.5 m

    d   20 m.

11. If the circumference of the bull's-eye measures 1.5 m, how many voussoirs will be needed to complete the bull's-eye? (Allow 10 mm for joints.)

    a   20

    b   25

    c   15

    d   18.

12. What is the area of a square that measures 5 m each side?

    a 0.25 m^2

    b 250 m^2

    c 25 m^2

    d 2.5 m^2.

13. What is the area of a rectangle measuring 7.5 m × 2.5 m?

    a 187.5 m^2

    b 18.75 m^2

    c 18.0 m^2

    d 1.85 m^2.

14. What is the area of a circle with a diameter of 2.5 m?

    a 4.5 m^2

    b 45 m^2

    c 50 m^2

    d 4.9 m^2.

15. What is the area of a circle with a radius of 0.9 m?

    a 25 m^2

    b 2.54 m^2

    c 2.0 m^2

    d 0.25 m^2.

16. What is the circumference of a circle with diameter of 4.5 m?

    a   142 m

    b   145 m

    c   14.13 m

    d   14.5 m.

17. What is the area of the triangle that has a base of 4.2 m and a vertical height of 2.0 m?

    a 4.5 m^2

    b 4.3 m^2

    c 4.2 m^2

    d 42 m^2.

18. What is the area of the triangle shown in the figure below?

  a   4 m²
  b   5 m²
  c   6 m²
  d   3 m².

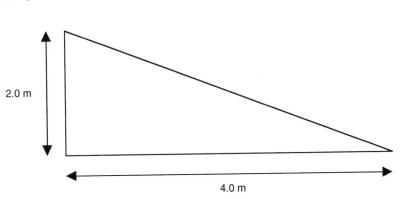

2.0 m

4.0 m

19. What is the circumference of a circle with a diameter of 1.75 m?

  a   5.5 m
  b   55 m
  c   50 m
  d   5.0 m.

20. What is the area of the triangle shown in the figure below?

  a   0.328 m²
  b   3.28 m²
  c   38 m²
  d   32 m².

1.75 m

3.75 m

21. What is the circumference of a semicircular arch with a span of 0.9 m?

  a   14.13 m
  b   15 m
  c   1.4 m
  d   1.5 m.

22. What is the volume of a path measuring 6.5 m × 1.5 m × 0.150 m?
    a   1.5 m³
    b   15 m³
    c   1.46 m³
    d   0.14 m³.

Now check your answers from the grid

Q 1; a	Q 6; a	Q 11; a	Q 16; c	Q 21; a
Q 2; d	Q 7; b	Q 12; c	Q 17; c	Q 22; c
Q 3; d	Q 8; c	Q 13; b	Q 18; a	
Q 4; d	Q 9; b	Q 14; d	Q 19; a	
Q 5; b	Q 10; a	Q 15; b	Q 20; b	

Section 8.2

1. If a door has a glass panel measuring 0.750 m × 0.500 m and the glass costs £1.50 per square metre, what is the cost of the glass?
    a   5.60p
    b   56p
    c   0.56p
    d   0.75p.

2. A window frame has 10 pieces of glass, each measuring 250 mm × 250 mm. What is the total area of glass in the frame?
    a   0.62 m²
    b   6.5 m²
    c   6.2 m²
    d   0.65 m².

3. A bathroom is to be half tiled with 150 mm × 150 mm tiles. If the area to be tiled is 3 m², how many tiles will be required to tile the bathroom?
    a   1.33
    b   133
    c   13
    d   150.

4. If a half-brick wall is 5 m × 2 m, how many bricks will be needed to build the wall?
    a   600
    b   60
    c   560
    d   56.

5. If a one-brick wall is 12 m × 3 m, how many bricks will be needed to build the wall?
  a 43.2
  b 432
  c 4320
  d 4500.

6. A wall is to be built in English bond 15 m × 2 m. How many bricks are needed to build the wall?
  a 3600
  b 360
  c 3500
  d 350.

7. If a wall is 5 m × 3.5 m, what is the total area?
  a 175 m²
  b 150 m²
  c 15 m²
  d 17.5 m².

8. How many square metres is in a wall 7.5 m × 3.5 m?
  a 26.25 m²
  b 265 m²
  c 25 m³
  d 2.5 m².

9. If a patio is 3.5 m × 2.5 m, what is the total area?
  a 87.5 m²
  b 58 m²
  c 8.75 m²
  d 58 m².

10. A window is 1.5 m × 0.9 m. What is the total area?
  a 137 m²
  b 13.5 m²
  c 1.35 m²
  d 1.5 m².

11. How many bricks will you need to build 15 m² of half-brick walling?
  a 1200
  b 900
  c 600
  d 1500.

12. If you are going to build a half-brick wall 9 m long by 2 m high, how many bricks are required?

    a   1080

    b   900

    c   1200

    d   1000.

13. How many bricks will be needed to build a 20 m² one-brick wall in Flemish bond?

    a   1200

    b   250

    c   2400

    d   120.

14. How many bricks will be needed to build a half-brick wall 15 m long by 3 m high?

    a   2700

    b   270

    c   240

    d   2400.

15. If you need 1 kg of mortar to lay one brick, how much mortar will be needed to lay 270 bricks?

    a   270 kg

    b   2700 kg

    c   27 kg

    d   2.7 kg.

16. How many bricks would be needed to build the half-brick wall shown in the figure?

    a   11.25

    b   1125

    c   500

    d   2500.

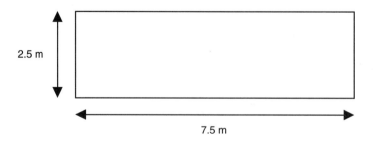

2.5 m

7.5 m

17. How many 0.6 m × 0.6 m paving flags are needed to pave a 15 m² patio?

    a   42

    b   426

    c   4.26

    d   50.

18. The figure shows a gable end of a garage wall built in 150 mm concrete blocks. How many blocks are needed to build the gable?

   a   34

   b   340

   c   43

   d   450.

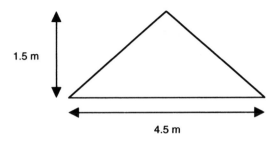

1.5 m

4.5 m

19. If the circumference of a semicircular arch measures 1.88 m, how many voussoirs will be needed to complete the arch?

   a   250

   b   50

   c   25

   d   500.

20. How much concrete is needed to lay a path measuring 9 m × 1.2 m × 0.150 m?

   a   1.62 m^3

   b   162 m^3

   c   16 m^3

   d   1.2 m^3.

21. How much concrete will be needed to lay a raft foundation 3 m × 4.5 m × 0.225 m?

   a   30 m^3

   b   3 m^3

   c   5 m^3

   d   4.5 m^3.

22. Calculate the volume of a foundation measuring 12.0 m × 0.75 m × 0.150 m.

   a   135 m^3

   b   150 m^3

   c   1.35 m^3

   d   1.50 m^3.

Now check your answers from the grid

Q 1; b	Q 6; a	Q 11; b	Q 16; b	Q 21; b
Q 2; a	Q 7; d	Q 12; a	Q 17; a	Q 22; c
Q 3; b	Q 8; a	Q 13; c	Q 18; a	
Q 4; a	Q 9; c	Q 14; a	Q 19; c	
Q 5; b	Q 10; c	Q 15; a	Q 20; a	

Section 8.3

Bricks

1. How many bricks are needed to build a half-brick wall in stretcher bond if the wall is 7.5 m long by 2.5 m high?
    (Add 5% for wastage)

2. How many bricks would be needed to build the wall shown in the figure?
    (Add 5% for wastage)

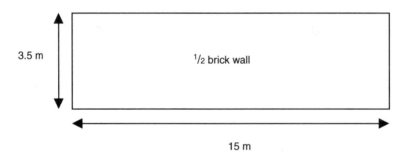

3. How many bricks will you need to build the wall shown below?

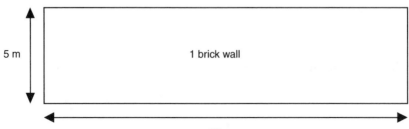

4. How many bricks are required to build the half-brick wall shown in the figure?

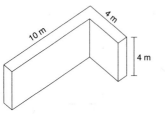

5. How many bricks will be needed to build the one-brick wall shown in the figure below?

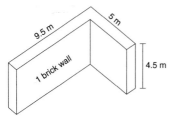

6. How many bricks will be needed to build the half-brick wall shown in the figure? (Add 5% for wastagage)

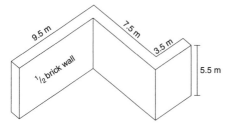

Blocks

7. How many standard size blocks will you need to build this wall?

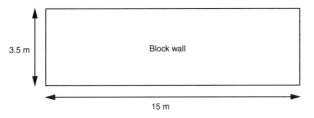

8. How many blocks will be required to build the wall shown below?

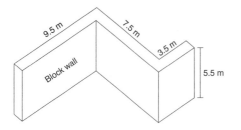

Mortar

Remember that it takes approximately 1 kg of mortar to lay one brick and 2 kg to lay one block.

9. Calculate the amount of bricks and mortar to build the wall shown.
    (Add 5% for wastage)

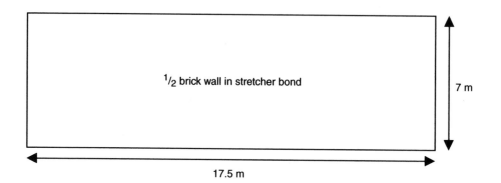

½ brick wall in stretcher bond

7 m

17.5 m

10. Calculate the number of bricks and the amount of mortar required to build the wall shown in the following figure.

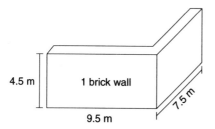

4.5 m

1 brick wall

9.5 m

7.5 m

11. How many standard blocks and how much mortar is needed to build the block wall shown in the figure?

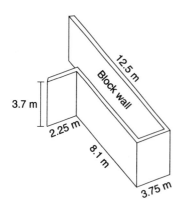

3.7 m

2.25 m

Block wall

12.5 m

8.1 m

3.75 m

Openings in walls

Remember that you must deduct any openings in the wall before working out the surface area.

12. How many bricks and how much mortar will be needed to build this wall.

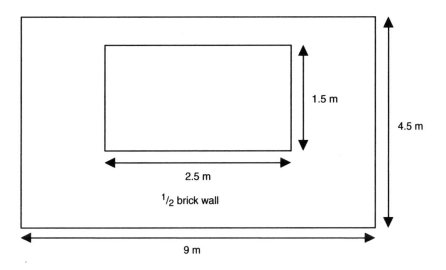

13. How many bricks and how much mortar is needed to build the wall shown in the following figure, assuming the building to be 12.8 m long by 3.75 m high?

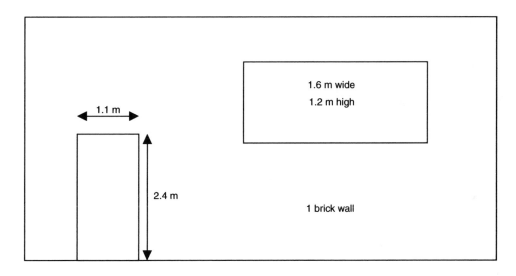

Cavity walls

You must work out each leaf of the cavity wall separately to achieve the totals before ordering bricks, blocks and mortar.

14. Calculate how many facing bricks, blocks and the amount of mortar needed to build a cavity wall 18 m long by 7.5 m high with an opening of 1.5 m by 2.4 m.

15. Calculate the quantities of facing bricks, blocks and mortar needed to build a cavity wall 16 m long by 8.5 m high, with two window openings of 1.5 m × 2.4 m and a door opening of 1.1 m × 2.4 m.

16. Calculate the number of facing bricks and blocks to build the cavity wall shown in the figure.

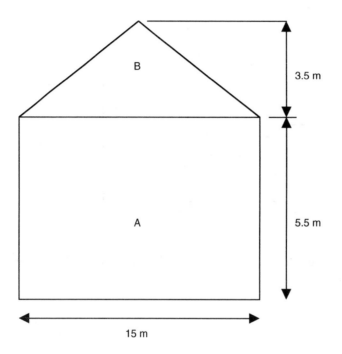

Division of facings and commons

When building walls one brick and over, it is sometimes cheaper to build the back of the wall with common bricks rather than use expensive facing bricks (see chart on page 158).

17. How many facings and common bricks will be needed to build a Flemish bond wall 25 m long by 2.5 m high?

18. How many facings and commons, and how much mortar will be needed to build an English bond wall 30 m long by 3.5 m high?

19. How many facings and common bricks will be needed to build an English garden wall if it is 18 m long by 3.5 m high?

Volumes

20. What volume of concrete is contained in an 8.4 m × 0.6 m × 0.225 m foundation?

21. What is the volume of concrete in a raft foundation measuring 5.5 m × 7.5 m × 0.300 m?

Areas of circles

22. Calculate the circular paved area shown in the figure.

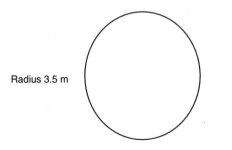

Radius 3.5 m

23. Calculate the shaded area of the bull's-eye.

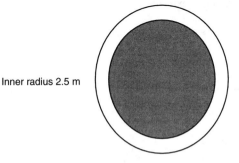

Inner radius 2.5 m

Estimating materials and costs

24. If a bricklayer's output is approximately 650 bricks laid per day, how long would you allow for two bricklayers to build a one-brick boundary wall, 125 m long and 1.75 m high?

25. Assume the bricks cost £120 per thousand and each bricklayer earns £10.50 per hour, and on average they work 8 hours per day. Including the cost of the bricks, how much would the wall in Question 24 cost to build?

Centre line method of measurement

26. Calculate the volume of earth excavated and the amount of concrete required for the foundation shown below. Use the centre line method.

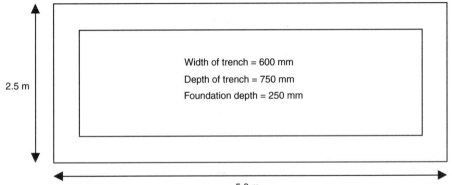

2.5 m

Width of trench = 600 mm
Depth of trench = 750 mm
Foundation depth = 250 mm

5.0 m

27. Using the centre line method, calculate the number of bricks and mortar to build the manhole shown below.

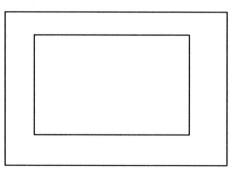 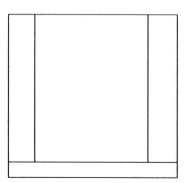

Size of manhole = 1.75 m × 1.5 m by 1.2 m high (external)

Model answers for Section 8.3

Bricks

1. How many bricks are needed to build a half-brick wall in stretcher bond if the wall is 7.5 m long by 2.5 m high?
(Add 5% for wastage)

$$\text{Area} = 7.5 \times 2.5 = 18.75 \text{ m}^2$$
$$\text{Bricks} = 18.75 \times 60 = 1125 + 5\% = 1181$$

2. How many bricks would be needed to build the wall shown in the figure?
(Add 5% for wastage)

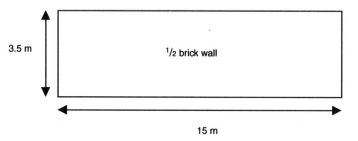

3.5 m

¹/₂ brick wall

15 m

$$\text{Area} = 15 \times 3.5 = 52.5 \text{ m}^2$$
$$\text{Bricks} = 52.5 \times 60 = 3150 + 5\% = 3307.5, \text{ say } 3310$$

3. How many bricks will you need to build the wall shown below?

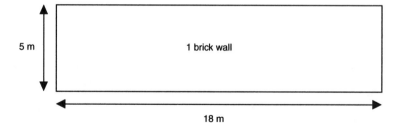

5 m

1 brick wall

18 m

Area = $18 \times 5 = 90 \text{ m}^2$
Bricks = $90 \times 120 = 10\ 800$ bricks

4. How many bricks are required to build the half-brick wall shown in the figure below?
Total length of wall = $10 + 4 = 14$ m

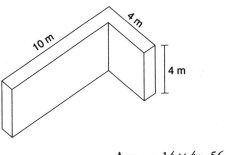

Area = $14 \times 4 = 56 \text{ m}^2$
Bricks = $56 \times 60 = 3360$

5. How many bricks will be needed to build the one-brick wall shown in the figure below?

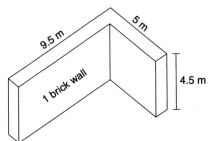

Total length of wall = $9.5 + 5 = 14.5$ m
Area = $14.5 \times 4.5 = 65.25 \text{ m}^2$
Bricks = $65.25 \times 120 = 7830$

6. How many bricks will be needed to build the half-brick wall shown in the figure?
(Add 5% for wastage)

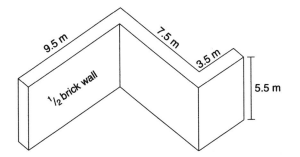

Total length of wall = $9.5 + 7.5 + 3.5 = 20.5$ m
Area = $20.5 \times 5.5 = 112.75 \text{ m}^2$
Bricks = $112.75 \times 60 = 6765$

Blocks

7. How many standard size blocks will you need to build this wall?

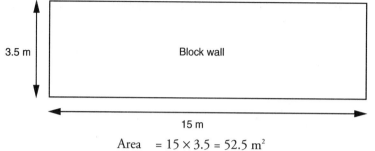

$$\text{Area} \quad = 15 \times 3.5 = 52.5 \text{ m}^2$$
$$\text{Blocks} = 52.5 \times 10 = 525$$

8. How many blocks will be required to build the wall shown below?

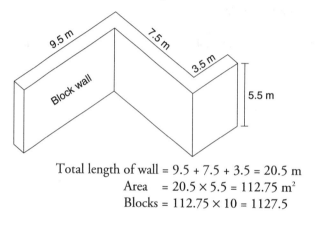

$$\text{Total length of wall} = 9.5 + 7.5 + 3.5 = 20.5 \text{ m}$$
$$\text{Area} \quad = 20.5 \times 5.5 = 112.75 \text{ m}^2$$
$$\text{Blocks} = 112.75 \times 10 = 1127.5$$

Mortar

Remember that it takes approximately 1 kg of mortar to lay one brick and 2 kg to lay one block.

9. Calculate the amount of bricks and mortar to build the wall shown.
(Add 5% for wastage)

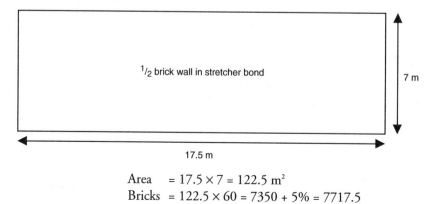

$$\text{Area} \quad = 17.5 \times 7 = 122.5 \text{ m}^2$$
$$\text{Bricks} = 122.5 \times 60 = 7350 + 5\% = 7717.5$$
$$\text{Mortar} = 7717.5 \text{ kg}$$

10. Calculate the number of bricks and the amount of mortar required to build the wall shown in the following figure.

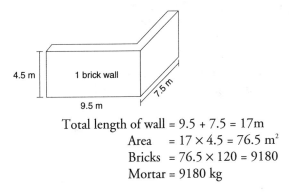

Total length of wall = 9.5 + 7.5 = 17m
Area   = 17 × 4.5 = 76.5 m²
Bricks = 76.5 × 120 = 9180
Mortar = 9180 kg

11. How many standard blocks and how much mortar is needed to build the block wall shown in the following figure?

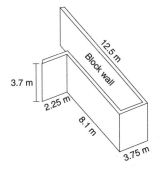

Total length of wall = 12.5 + 3.75 + 8.1 + 2.25 = 26.6 m
Area   = 26.6 × 3.7 = 98.42 m²
Blocks = 98.42 × 10 = 984.2
Mortar = 984.2 × 2 = 1968.4 kg

Openings in walls

Remember that you must deduct any openings in the wall before working out the surface area.

12. How many bricks and how much mortar will be needed to build this wall?

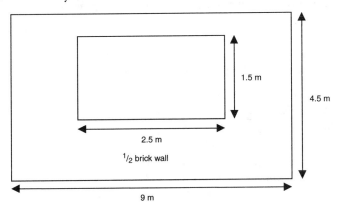

$$\text{Total area} = 9 \times 4.5 = 40.5 \text{ m}^2$$
$$\text{Window} = 2.5 \times 1.5 = 3.75 \text{ m}^2$$
$$\text{Surface area of brickwork} = 40.5 - 3.75 = 36.75 \text{ m}^2$$
$$\text{Bricks} = 36.75 \times 60 = 2205$$
$$\text{Mortar} = 2205 \times 1 = 2205 \text{ kg}$$

13. How many bricks and how much mortar is needed to build the wall shown in the following figure, assuming the building to be 12.8 m long by 3.75 m high?

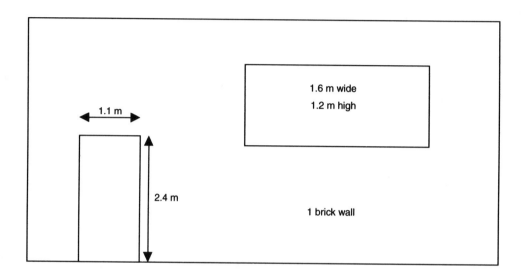

$$\text{Total area of wall} = 12.8 \times 3.75 = 48 \text{ m}^2$$
$$\text{Area of window} = 1.6 \times 1.2 = 1.92 \text{ m}^2$$
$$\text{Area of door} = 2.4 \times 1.1 = 2.64 \text{ m}^2$$
$$\text{Surface area of brickwork} = 48 - (1.92 + 2.64)$$
$$= 48 - 4.56 = 43.44 \text{ m}^2$$
$$\text{Bricks} = 43.44 \times 120 = 5212.8, \text{ say } 5213$$
$$\text{Mortar} = 5213 \times 1 = 5213 \text{ kg}$$

Cavity walls

You must work out each leaf of the cavity wall separately to achieve the totals before ordering bricks, blocks and mortar.

14. Calculate how many facing bricks, blocks and the amount of mortar needed to build a cavity wall 18 m long by 7.5 m high with an opening of 1.5 m by 2.4 m.

$$\text{Total area} = 18 \times 7.5 = 135 \text{ m}^2$$
$$\text{Area of opening} = 2.4 \times 1.5 = 3.6 \text{ m}^2$$
$$\text{Surface area of brick/blockwork} = 135 - 3.6 = 131.4 \text{ m}^2$$
$$\text{Facing bricks} = 131.4 \times 60 = 7884$$
$$\text{Blocks} = 131.4 \times 10 = 1314$$
$$\text{Mortar} = (7884 \times 1) + (1314 \times 2)$$
$$= 7884 + 2628 = 10\,512 \text{ kg}$$

15. Calculate the quantities of facing bricks, blocks and mortar needed to build a cavity wall 16 m long by 8.5 m high, with two window openings 1.5 m × 2.4 m and a door opening of 1.1 m × 2.4 m

$$\text{Total area} = 16 \times 8.5 = 136 \text{ m}^2$$
$$\text{Area of window openings} = 2.4 \times 1.5 \times 2 = 7.2 \text{ m}^2$$
$$\text{Door opening} = 2.4 \times 1.1 = 2.64 \text{ m}^2$$
$$\text{Surface area} = 136 - (7.2 + 2.64)$$
$$= 136 - 9.84 = 126.16 \text{ m}^2$$
$$\text{Facing bricks} = 126.16 \times 60 = 7569.6, \text{ say } 7570$$
$$\text{Blocks} = 126.16 \times 10 = 1261.6, \text{ say } 1262$$
$$\text{Mortar} = (7570 \times 1) + (1262 \times 2)$$
$$= 7570 + 2524 = 10\ 094 \text{ kg}$$

16. Calculate the number of facing bricks and blocks to build the cavity wall shown in the figure.

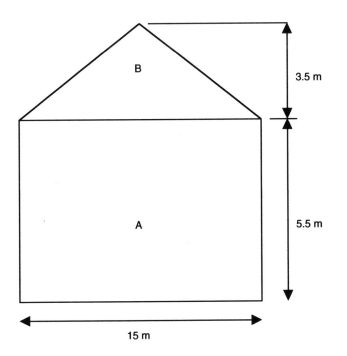

$$\text{Area of A} = 15 \times 5.5 = 82.5 \text{ m}^2$$
$$\text{Area of B} = \frac{15 \times 3.5}{2}$$
$$= 26.25 \text{ m}^2$$
$$\text{Total area} = 82.5 + 26.25 = 108.75 \text{ m}^2$$
$$\text{Number of facing bricks} = 108.75 \times 60 = 6525$$
$$\text{Number of blocks} = 108.75 \times 10 = 1087.5$$

Division of facings and commons

When building walls one brick and over, it is sometimes cheaper to build the back of the wall with common bricks rather than use expensive facing bricks (see chart on page 158).

17. How many facings and common bricks will be needed to build a Flemish bond wall 25 m long by 2.5 m high?

$$\text{Area} = 25 \times 2.5 = 62.5 \text{ m}^2$$
$$\text{Number of facings} = 62.5 \times 80 = 5000$$
$$\text{Number of commons} = 62.5 \times 40 = 2500$$

18. How many facings and commons, and how much mortar will be needed to build an English bond wall 30 m long by 3.5 m high?

$$\text{Area of wall} = 30 \times 3.5 = 105 \text{ m}^2$$
$$\text{Number of facings} = 105 \times 90 = 9450$$
$$\text{Number of commons} = 105 \times 30 = 3150$$
$$\text{Amount of mortar} = 9450 + 3150 = 12\ 600 \text{ kg}$$

19. How many facings and common bricks will be needed to build an English garden wall if it is 18 m long by 3.5 m high?

$$\text{Area of wall} = 18 \times 3.5 = 63 \text{ m}^2$$
$$\text{Number of facings} = 63 \times 73 = 4599$$
$$\text{Number of commons} = 63 \times 47 = 2961$$

Volumes

20. What volume of concrete is contained in an 8.4 m × 0.6 m × 0.225 m foundation?

$$\text{Volume} = 8.4 \times 0.6 \times 0.225 = 1.134 \text{ m}^3$$

21. What is the volume of concrete in a raft foundation measuring 5.5 m × 7.5 m × 0.300 m?

$$\text{Volume} = 5.5 \times 7.5 \times 0.3 = 12.375 \text{ m}^3$$

Areas of circles

22. Calculate the circular paved area shown in the figure.

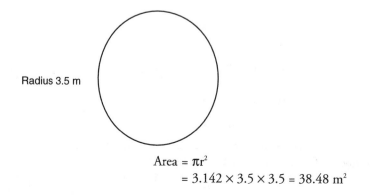

Radius 3.5 m

$$\text{Area} = \pi r^2$$
$$= 3.142 \times 3.5 \times 3.5 = 38.48 \text{ m}^2$$

23. Calculate the shaded area of the bull's-eye.

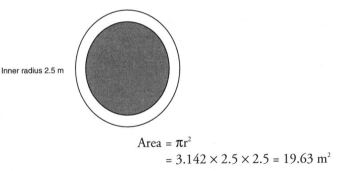

Inner radius 2.5 m

$$Area = \pi r^2$$
$$= 3.142 \times 2.5 \times 2.5 = 19.63 \text{ m}^2$$

Estimating materials and costs

24. If a bricklayer's output is approximately 650 bricks laid per day, how long would you allow for two bricklayers to build a one-brick boundary wall, 125 m long and 1.75 m high?

$$Area = 125 \times 1.75 = 218.75 \text{ m}^2$$
$$Number \ of \ bricks = 218.75 \times 120 = 26\ 250$$

One bricklayer lays 650, therefore two bricklayers will lay 1300 per day

$$Number \ of \ days = \frac{26\ 250}{1300}$$
$$= 20.19, \text{ say 21 days}$$

25. Assume the bricks cost £120 per thousand and each bricklayer earns £10.50 per hour, and on average they work 8 hours per day. Including the cost of the bricks, how much would the wall in question 24 cost to build?

$$Cost \ of \ bricks = \frac{26\ 250 \times 120}{1000}$$
$$= 3150$$
$$Hours \ worked = 8 \times 2 \times 21 = 336$$
$$Cost \ of \ labour = 336 \times £10.50 = £3528$$
$$Total \ cost = 3150 + £3528 = £6678$$

Centre line method measurement

26. Calculate the volume of earth excavated and the amount of concrete required for the foundation shown below. Use the centre line method.

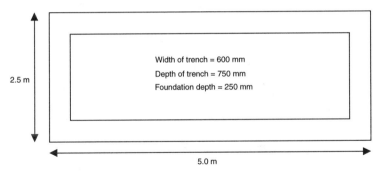

Width of trench = 600 mm
Depth of trench = 750 mm
Foundation depth = 250 mm

2.5 m

5.0 m

$$\text{Centre line} = (2.5 \times 2) + (5 \times 2) - (4 \times 0.600)$$
$$= 5 + 10 - 2.4 = 12.6$$
$$\text{Amount of earth excavated} = 12.6 \times 0.75 = 9.45 \text{ m}^3$$
$$\text{Volume of concrete required} = 12.6 \times 0.6 \times 0.250 = 1.89 \text{ m}^3$$

27. Using the centre line method, calculate the number of bricks and mortar to build the manhole shown below.

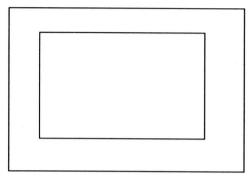 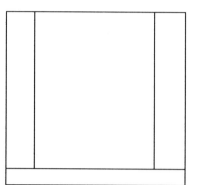

Size of manhole = 1.75 m × 1.5 m by 1.2 m high (external)

$$\text{Centre line} = (1.75 \times 2) + (1.5 \times 2) - (4 \times 0.225)$$
$$= 3.5 + 3.0 - 0.9 = 5.6 \text{ m}$$
$$\text{Area} = 5.6 \times 1.2 = 6.72 \text{ m}^2$$
$$\text{No. of bricks} = 6.72 \times 120 = 806.4, \text{ say 810 bricks}$$
$$\text{Mortar} = 810 \times 1 \text{ kg} = 810 \text{ kg}$$

NOTES